Asymptotic Expansions for Infinite Weighted Convolutions of Heavy Tail Distributions and Applications

of the
American Mathematical Society

Number 922

Asymptotic Expansions for Infinite Weighted Convolutions of Heavy Tail Distributions and Applications

Ph. Barbe
W.P. McCormick

January 2009 • Volume 197 • Number 922 (fourth of 5 numbers) • ISSN 0065-9266

American Mathematical Society
Providence, Rhode Island

2000 *Mathematics Subject Classification.* Primary 41A60, 60F99;
Secondary 41A80, 44A35, 60E07, 60G50, 60K05, 60K25, 62E17, 62G32.

Library of Congress Cataloging-in-Publication Data

Barbe, Philippe.

Asymptotic expansions for infinite weighted convolutions of heavy tail distributions and applications / Ph. Barbe, W. P. McCormick.

    p. cm. — (Memoirs of the American Mathematical Society, ISSN 0065-9266 ; no. 922)

"Volume 197, number 922 (Fourth of five numbers)."

Includes bibliographical references and index.

ISBN 978-0-8218-4259-1 (alk. paper)

    1. Distribution (Probability theory)—Mathematical models. 2. Asymptotic expansions. 3. Stochastic processes. I. McCormick, William P.

QA273.6.B376   2009

519.2′4—dc22
                                                                                                                2008039491

Memoirs of the American Mathematical Society

This journal is devoted entirely to research in pure and applied mathematics.

Subscription information. The 2009 subscription begins with volume 197 and consists of six mailings, each containing one or more numbers. Subscription prices for 2009 are US$709 list, US$567 institutional member. A late charge of 10% of the subscription price will be imposed on orders received from nonmembers after January 1 of the subscription year. Subscribers outside the United States and India must pay a postage surcharge of US$65; subscribers in India must pay a postage surcharge of US$95. Expedited delivery to destinations in North America US$57; elsewhere US$160. Each number may be ordered separately; *please specify number* when ordering an individual number. For prices and titles of recently released numbers, see the New Publications sections of the *Notices of the American Mathematical Society*.

Back number information. For back issues see the *AMS Catalog of Publications*.

Subscriptions and orders should be addressed to the American Mathematical Society, P. O. Box 845904, Boston, MA 02284-5904, USA. *All orders must be accompanied by payment*. Other correspondence should be addressed to 201 Charles Street, Providence, RI 02904-2294, USA.

Copying and reprinting. Individual readers of this publication, and nonprofit libraries acting for them, are permitted to make fair use of the material, such as to copy a chapter for use in teaching or research. Permission is granted to quote brief passages from this publication in reviews, provided the customary acknowledgment of the source is given.

Republication, systematic copying, or multiple reproduction of any material in this publication is permitted only under license from the American Mathematical Society. Requests for such permission should be addressed to the Acquisitions Department, American Mathematical Society, 201 Charles Street, Providence, Rhode Island 02904-2294, USA. Requests can also be made by e-mail to reprint-permission@ams.org.

Memoirs of the American Mathematical Society (ISSN 0065-9266) is published bimonthly (each volume consisting usually of more than one number) by the American Mathematical Society at 201 Charles Street, Providence, RI 02904-2294, USA. Periodicals postage paid at Providence, RI. Postmaster: Send address changes to Memoirs, American Mathematical Society, 201 Charles Street, Providence, RI 02904-2294, USA.

© 2009 by the American Mathematical Society. All rights reserved.
Copyright of individual articles may revert to the public domain 28 years
after publication. Contact the AMS for copyright status of individual articles.
This publication is indexed in *Science Citation Index*®, *SciSearch*®, *Research Alert*®,
CompuMath Citation Index®, *Current Contents*®/*Physical, Chemical & Earth Sciences*.
Printed in the United States of America.

∞ The paper used in this book is acid-free and falls within the guidelines
established to ensure permanence and durability.
Visit the AMS home page at http://www.ams.org/

10 9 8 7 6 5 4 3 2 1   14 13 12 11 10 09

Contents

1. Introduction — 1
   1.1. Prolegomenom — 1
   1.2. Mathematical overview and heuristics — 4

2. Main result — 9
   2.1. Some notation — 9
   2.2. Asymptotic scales — 10
   2.3. The Laplace characters — 12
   2.4. Smoothly varying functions of finite order — 15
   2.5. Asymptotic expansion for infinite weighted convolution — 16

3. Implementing the expansion — 21
   3.1. How many terms are in the expansion? — 21
   3.2. $\star$-Asymptotic scales and functions of class m — 24
   3.3. Tail calculus: From Laplace characters to linear algebra — 27
   3.4. Some examples — 28
   3.5. Two terms expansion and second order regular variation — 34
   3.6. Some open questions — 36

4. Applications — 39
   4.1. ARMA models — 39
   4.2. Tail index estimation — 40
   4.3. Randomly weighted sums — 47
   4.4. Compound sums — 50
   4.5. Queueing theory — 53
   4.6. Branching processes — 55
   4.7. Infinitely divisible distributions — 56
   4.8. Implicit transient renewal equation and iterative systems — 58

5. Preparing the proof — 65
   5.1. Properties of Laplace characters — 65
   5.2. Properties of smoothly varying functions of finite order — 67

6. Proof in the positive case — 75
   6.1. Decomposition of the convolution into integral and multiplication operators — 75
   6.2. Organizing the proof — 77
   6.3. Regular variation and basic tail estimates — 79
   6.4. The fundamental estimate — 82
   6.5. Basic lemmas — 83
   6.6. Inductions — 89
   6.7. Conclusion — 94

7. Removing the sign restriction on the random variables   97
    7.1. Elementary properties of U_H 98
    7.2. Basic expansion of U_H 99
    7.3. A technical lemma   100
    7.4. Conditional expansion and removing conditioning   102

8. Removing the sign restriction on the constants   105
    8.1. Neglecting terms involving the multiplication operators   105
    8.2. Substituting $\overline{H}^{(k)}$ and $\overline{G}^{(k)}$ by their expansions   107

9. Removing the smoothness restriction   109

Appendix. `Maple` code   111

Bibliography   115

Abstract

We establish some asymptotic expansions for infinite weighted convolution of distributions having regularly varying tails. Applications to linear time series models, tail index estimation, compound sums, queueing theory, branching processes, infinitely divisible distributions and implicit transient renewal equations are given.

A noteworthy feature of the approach taken in this paper is that through the introduction of objects, which we call the Laplace characters, a link is established between tail area expansions and algebra. By virtue of this representation approach, a unified method to establish expansions across a variety of problems is presented and, moreover, the method can be easily programmed so that a computer algebra package makes implementation of the method not only feasible but simple.

Received by the editor February 18, 2006, and in revised form August 23, 2006.

2000 Mathematics Subject Classification. Primary 41A60, 60F99; secondary 41A80, 44A35, 60E07, 60G50, 60K05, 60K25, 62E17, 62G32.

Key words. asymptotic expansion, regular variation, convolution, tail area approximation, ARMA models, tail estimation, compound sums, infinitely divisible distributions, renewal theory.

Acknowlegements. Philippe Barbe thanks Peter Haskell and Konstantin Mischaikow for changing his views on algebraic constructions through their marvelous teaching.

CHAPTER 1

Introduction.

The primary focus of this paper is to obtain precise understanding of the distribution tail of linear and related stochastic processes based on heavy tail innovations. In doing so, we will develop some new mathematical objects which are tailored to efficiently write and compute asymptotic expansions of these tails. Also, we will derive simple bounds of theoretical importance for the error between the tail and its asymptotic expansion.

These tails and their expansions are of interest in a variety of contexts. In the following subsection, we provide some typical examples to illustrate their use. The second subsection of this introduction overviews the new perspective and techniques developed in this paper; this will be done at a heuristic level, explaining the intuition and sketching the broad expanse wherein our methods lie.

1.1. Prolegomenom.

The first basic problem we will deal with is related to the tail behavior of the marginal distribution of linear processes. To be specific, let $C = (c_i)_{i \in \mathbb{Z}}$ be a sequence of real constants, and let $X = (X_i)_{i \in \mathbb{Z}}$ be a sequence of independent and identically distributed random variables. Let F be the common distribution function of these X_i's and write $\overline{F} = 1 - F$ for the tail function. Assume that F has a heavy tail, that is, there exists a positive and finite α such that for any positive λ,

$$\lim_{t \to \infty} \overline{F}(t\lambda)/\overline{F}(t) = \lambda^{-\alpha}. \tag{1.1.1}$$

Given the sequence C and the distribution function F, we write F_C for the distribution function of the series $\sum_{i \in \mathbb{Z}} c_i X_i$. Set $\overline{F}_C = 1 - F_C$. For simplicity, assume in this introduction that the sequence C is nonnegative. Define the series $C_\alpha = \sum_{i \in \mathbb{Z}} c_i^\alpha$. It is well known that under a mild additional condition,

$$\overline{F}_C \sim C_\alpha \overline{F}$$

at infinity, that is $\lim_{t \to \infty} \overline{F}_C(t)/\overline{F}(t) = C_\alpha$.

It is then a natural question to investigate higher-order expansions. Under suitable conditions, we will obtain higher-order asymptotic expansions both for $\overline{F}_C$ and its derivatives. In particular all ARMA processes fall within the scope of our result, provided the innovation distribution satisfies certain mild conditions beyond that of heavy tail.

This question of obtaining higher-order terms has several motivations. One is that very little is known on the marginal distributions of linear processes, and since those are ubiquitous in statistics, better knowledge and understanding of their properties is desirable. There are also specific instances where higher-order asymptotics are needed, such as in the tail index estimation problem which we will study in section 4.2. To achieve such refined distributional results in this setting, we must first build the mathematical language and theory needed for higher-order expansions.

A third motivation comes from a large number of applications to related processes obtained by allowing the weights to be random. Whereas many first-order results are known, few second-order results exist, and no higher-order results are available that we are aware of. We will obtain asymptotic expansions when the weights are random, not necessarily independent or identically distributed, but still assuming that the sequence C, now random, is independent of X.

This will allow us to derive tail expansions for compound sums, that is a sum with a random number of summands. As a consequence, we will derive a tail expansion for some infinitely divisible distributions. In the same vein, having a result with random weights yields expansions for tails arising in transient renewal theory, and even in implicit transient renewal theory. By the latter, we mean, for example, distributions defined implicitly as follows. Let M, Q and R be independent nonnegative random variables and suppose that R has the same distribution as $Q + MR$. Given the distributions of M and Q, this defines that of R implicitly. We will obtain an asymptotic expansion for the distribution tail of R, when Q has a heavy tail and M has enough moments.

Furthermore, it will be clear that by the algebraic paradigm of this paper, it is quite a simple matter to write out what the expansions should be, and often not so difficult to prove that those expansions are valid under some reasonable conditions.

In all the contexts mentioned, first-order results exist under heavy tail assumptions, or more generally in the subexponential framework. By now, these results are well understood. We refer to the book by Bingham, Goldie and Teugels (1989) for a comprehensive study of these topics and Broniatowski and Fuchs (1995) for a different perspective on first-order results for sums of independent and identically distributed random variables. Resnick (1986) is a good reference for many first-order results obtained through a point process argument. In comparison, second-order results are few. In the heavy-tail case papers which provide second-order results for the sum of two random variables include Omey (1988), Geluk (1994), Geluk, De Haan, Resnick and Stărică (1997) and Geluk, Peng and De Vries (2000). With regard to subexponential distributions, second-order work for a convolution of two such distributions may be found in Cline (1986, 1987) and Geluk and Pakes (1991). We also mention that Geluk, Peng and De Vries (2000) shows a second-order result for a sum of a finite number of independent and identically distributed random variables in the special case of the underlying distribution being a member of the Hall-Weissman class. Barbe and McCormick (2005) obtained second-order results for the sum of a finite number of heavy tailed random variables when the distributions also possess a mild smoothness property. Using a different set of assumptions and for a different purpose, Borovkov and Borovkov (2003) obtained higher-order expansions also for a finite number of summands.

In renewal theory, a little more is known. For instance, in recurrent renewal theory, higher-order results are already in Feller (1971). But our technique is useful only in the transient case, for which far less is known. Chover, Ney and Wainger (1973) provides a first-order result while Grübel (1987) gives a second-order formula. In the implicit case, Goldie (1991) obtained second-order formulas. However, our work complements nicely with Goldie's, in the sense that our results typically apply when his fail. The techniques used by Goldie (1991) build on those devised by Kesten (1973) and Grincevičius (1975), and make use of delicate Tauberian theory, while that of Grübel (1987) falls under the so-called Banach algebra technique. Our method is radically different from these.

Grübel's (1987) result on renewal measures allows him to obtain a second-order formula for the tail of some infinitely divisible distributions, using a decomposition which we could trace back at least to Feller (1971). We will use the same decomposition to obtain higher-order results.

Since our approach to deriving these asymptotic expansions may appear very algebraic to analytically oriented minds, we feel compelled at this point to quote a result which can be proved in less than 5 minutes after understanding the statement of our main result (Theorem 2.3.1) and being familiar with our algebraic formalism. Consider the distribution function whose tail is $\overline{F}(t) = (a+b)^{-1}(at^{-\alpha} + bt^{-\beta})$ for t more than 1, and with $1 < \alpha < \beta$. Assume further that the c_i's are nonnegative. Write $C_r = \sum_{i \in \mathbb{Z}} c_i^r$, and

$$\mu_{F,1} = \frac{1}{a+b}\left(\frac{a}{\alpha-1} + \frac{b}{\beta-1}\right)$$

for the first moment of F. Then we have the two terms expansion

$$\overline{F}_C(t) = \frac{a}{a+b}\frac{C_\alpha}{t^\alpha}$$

$$+ \begin{cases} \dfrac{bC_\beta}{a+b}\dfrac{1}{t^\beta} & \text{if } \beta < \alpha+1 \\ \dfrac{1}{a+b}\bigl(bC_\beta + a\mu_{F,1}\alpha(C_1 C_\alpha - C_{\alpha+1})\bigr)\dfrac{1}{t^{\alpha+1}} & \text{if } \beta = \alpha+1 \\ \dfrac{a\alpha}{a+b}\mu_{F,1}(C_1 C_\alpha - C_{\alpha+1})\dfrac{1}{t^{\alpha+1}} & \text{if } \alpha+1 < \beta \end{cases} + o\left(\frac{1}{t^{\alpha+1}}\right).$$

With our formalism and notations to be explained later, this complicated formula has the far more manageable form, where no cases need to be distinguished,

$$\overline{F}_C \sim \sum_{c \in C} L_{F_{C \setminus c}, 1} \overline{M_c F}.$$

Note that if β is $\alpha + 1$, the c_i's and b can be chosen so that the above second-order expansion yields only

$$\overline{F}_C(t) = \frac{a}{a+b} C_\alpha t^{-\alpha} + o(t^{-\alpha-1}).$$

It is then desirable to get an extra term. In section 3.1, we will obtain this term, generically, when α is more than 2. We will also explain why in some very exceptional cases, obtaining a second term is rather hopeless (last statement of section 3.1), while in other cases, obtaining as many terms as one desires can be trivial (penultimate example in section 3.4, dealing with the log-gamma distribution).

To conclude this preliminary discussion, and perhaps to motivate further our investigation, we mention that there are many unsolved very basic problems related to linear processes. For instance, for discrete innovations, Davis and Rosenblatt (1991) shows that all but trivial infinite-order linear processes have a continuous marginal distribution; but there is still no good criterion to determine if the marginal distribution of the process is absolutely continuous. Even in the most basic cases, the behavior of the marginal distribution is amazingly difficult to analyze. Solomyak's (1995) breakthrough — a generic result on absolute continuity of the marginal distribution of first-order autoregressive models with Bernoulli innovations — gives a stark reminder on how little is known in general. So this paper can be taken as a contribution to the understanding of those distributions, in the continuous and heavy tail situation.

1.2. Mathematical overview and heuristics.

We now outline the mathematical content of the key parts of this paper and provide some intuition behind the main result.

In our opinion, the main reason why so few higher-order results are available in the problems we are interested in, is that too much emphasis has been given to an analytical perspective. Our first basic remark is that the convolution operation $(F, G) \mapsto F \star G$ is bilinear and defines a semi-group. Bilinearity is a notion in linear algebra; while semi-group is related to the group structure, which in turn suggests representation theory. So our view is that the asymptotic behavior of the convolution semi-group should be analyzed by a linear representation which captures only the tail behavior of the semi-group. In fact we will obtain two linear representations, one which is suitable for writing the expansions, the other one, derived from the first one, and requiring more assumptions, suitable for practical computations.

To explain further our view on the subject, let us give a heuristic argument on how to derive higher-order expansions and build the mathematical language to handle these expansions.

Heuristic. Let F and G be two distribution functions. Their convolution is

$$\overline{F \star G}(t) = \int \overline{F}(t-x) \, dG(x).$$

The left hand side in this formula is symmetric, while the right hand side, as far as the notation goes, is not. To symmetrize it, we split the integral and integrate by parts to obtain

$$\overline{F \star G}(t) = \int_{-\infty}^{t/2} \overline{F}(t-x) \, dG(x) + \int_{-\infty}^{t/2} \overline{G}(t-x) \, dF(x) + \overline{F}(t/2)\overline{G}(t/2).$$

This formula involves the translation τ_x acting on functions and defined by $\tau_x h(t) = h(t+x)$. The translations form a semigroup since $\tau_x \tau_y = \tau_{x+y}$. The corresponding infinitesimal generator is the derivative D, because for smooth functions

$$\lim_{\epsilon \to 0} \frac{\tau_\epsilon - \tau_0}{\epsilon} h(t) = \lim_{\epsilon \to 0} \frac{h(t+\epsilon) - h(t)}{\epsilon} = \mathrm{D}h(t).$$

So, one expects to be able to write $\tau_x = e^{x\mathrm{D}}$ — this is nothing but a neat way to write a Taylor expansion, which we could trace back to Delsarte (1938). Indeed, applying both sides to a smooth function h yields

$$h(t+x) = \tau_x h(t) = e^{x\mathrm{D}} h(t) = \sum_{i \geqslant 0} \frac{x^i \mathrm{D}^i}{i!} h(t) = \sum_{i \geqslant 0} \frac{x^i}{i!} h^{(i)}(t).$$

So, formally,

$$\int_{-\infty}^{t/2} \overline{F}(t-x)\,\mathrm{d}G(x) = \int_{-\infty}^{t/2} \tau_{-x}\,\mathrm{d}G(x)\overline{F}(t) = \int_{-\infty}^{t/2} e^{-x\mathrm{D}}\,\mathrm{d}G(x)\overline{F}(t).$$

As t tends to infinity, the linear operator $\int_{-\infty}^{t/2} e^{-x\mathrm{D}}\,\mathrm{d}G(x)$ should tend to the Laplace transform of G at D, say $L_G(\mathrm{D})$. So, ideally, we have

$$\overline{F \star G} = L_G(\mathrm{D})\overline{F} + L_F(\mathrm{D})\overline{G} + O(\overline{F}\,\overline{G}). \tag{1.2.1}$$

Of course, for heavy tail distributions, many things are not well defined. Certainly, the Laplace transform does not exist, the distribution lacking sufficient integrability.

So, consider the Taylor polynomial of the Laplace transform. Write $\mu_{G,k}$ for the k-th moment of G. Assume that these moments are finite for k at most equal to some m. The Taylor polynomial of the Laplace transform gives rise to the differential operator

$$L_{G,m} = \sum_{0 \leqslant j \leqslant m} \frac{(-1)^j}{j!} \mu_{G,j} \mathrm{D}^j.$$

Assume that $\overline{F}$ and $\overline{G}$ are regularly varying of index $-\alpha$, that is obey (1.1.1). Take m be an integer less than α. Throughout this paper, Id denotes the identity function of whatever space is under consideration. When using the identity on the real line, we write Id^k for the function $t \mapsto t^k$. So, $\mathrm{Id}^{-m}\overline{F}$ is the function whose value at t is $t^{-m}\overline{F}(t)$. We certainly would like to have the m-th derivative $\overline{F}^{(m)}$ regularly varying of index $-\alpha - m$. In that case $(L_G(\mathrm{D}) - L_{G,m})\overline{F}$ could be of order smaller than the last term of $L_{G,m}\overline{F}$, that is $\overline{F}^{(m)}$, or equivalently, $\mathrm{Id}^{-m}\overline{F}$, which in turn dominates $\overline{F}\,\overline{G}$. So replacing $L_G(\mathrm{D})$ and $L_F(\mathrm{D})$ in (1.2.1) by $L_{G,m}$ and $L_{F,m}$ and taking m less than α suggests

$$\overline{F \star G} = L_{G,m}\overline{F} + L_{F,m}\overline{G} + o(\mathrm{Id}^{-m}\overline{F}). \tag{1.2.2}$$

The advantage of this substitution is that now everything is well defined, in particular it does not assume that F and G have moments of arbitrary large order.

By induction, if we have n distribution functions F_i, $1 \leqslant i \leqslant n$, we obtain

$$\overline{\star_{1\leqslant i\leqslant n}F_i} = \sum_{1\leqslant i\leqslant n} L_{\star_{\substack{1\leqslant j\leqslant n \\ j\neq i}} F_j, m} \overline{F_i} + \text{remainder}.$$

Taking limit as n tends to infinity yields a formula, which, restricted to the context of linear processes, is in fact the main result of this paper.

The substantial task of transforming this heuristic argument into a rigorous proof will be carried out in sections 5, 6, 7, 8 and 9.

Going back to our heuristic, where is the representation which we announced? We will show that in the proper ring, that of polynomials in D modulo the ideal generated by D^{m+1}, the composition of $L_{F,m}$ with $L_{G,m}$ gives $L_{F\star G,m}$. So the map $F \mapsto L_{F,m}$ turns out to be a linear representation of the convolution algebra in a ring of differential operators. In this ring, the operators $L_{F,m}$ will be invertible.

This gives us a very powerful formalism to state expansions. But, arguably, it is not quite explicit. To obtain a powerful computational machinery, we need to go from a ring of differential operators to a vector space where we can manipulate finite dimensional matrices. This is not always possible though, and it is related to a rather intricate matter concerning asymptotic scales. Those will be introduced in subsection 2.2 and further examined in our context in subsection 3.2.

Nevertheless, the basic idea is quite intuitive. For any positive real number c, write $M_c F$ to denote the distribution function of cX_i. More generally, when c is positive and h is a function, we write $M_c h$ for the function whose value at t is $h(t/c)$. Since $\mathrm{D}\overline{M_c F} = c^{-1} M_c \mathrm{D}\overline{F}$, the expansion (1.2.2) with $m = 1$ yields

$$\overline{M_{c_1} F \star M_{c_2} F} = M_{c_1}\overline{F} + M_{c_2}\overline{F} - c_1 \mu_{F,1} c_2^{-1} M_{c_2} \mathrm{D}\overline{F}$$
$$- c_2 \mu_{F,1} c_1^{-1} M_{c_1} \mathrm{D}\overline{F} + \text{remainder}. \qquad (1.2.3)$$

This suggests that we could think of such expansion as a decomposition in a basis comprised of $\overline{F}$ and $\mathrm{D}\overline{F}$, that is as a sort of projection onto the 2-dimensional vector space spanned by these functions. But there is a problem to overcome, namely that the sum $M_{c_1}\overline{F} + M_{c_2}\overline{F}$ is not quite the usual first-order result $(c_1^\alpha + c_2^\alpha)\overline{F}$, except if both c_1 and c_2 are 1 or $\overline{F}$ is exactly a Pareto distribution. In doing such a replacement, we make an error, which may be larger than the second-order term. So, even in an asymptotic sense, $M_c\overline{F}$ may not be in the space spanned by $\overline{F}$ and $\mathrm{D}\overline{F}$. We will give examples in sections 3.1 and 3.4 showing that this problem is not a failure of the expansion, but that of the basis $\overline{F}$, $\mathrm{D}\overline{F}$. This suggests that some bases are better than others, and of course raises the question of what is a good basis for our purpose. It also suggests that when all the weights are 1, the natural basis may be that of $\overline{F}$ and its derivatives.

Since we will now be dealing with issues related to asymptotic analysis, we will switch to a terminology used in that field. Specifically, we will refer to an asymptotic scale for that which we called a basis in the previous paragraph. Asymptotic scales will be introduced formally in subsection 2.2.

In studying linear processes, two operations are involved. One is the convolution obtained from summing independent random variables, the other is the scaling M_c obtained by rescaling the variables. We have seen that as far as the convolution goes, we have a representation using the differential operators $L_{F,m}$, which themselves involve D. This suggests that the good asymptotic scales are in some sense stable by differentiation and rescaling. In which sense? Since we are interested in asymptotic expansions, and we want to preserve the linear aspect of the convolution, they should be stable in the sense that derivatives and scaling of functions in this scale should have an asymptotic expansion in the same scale. The simplest example of such functions are the powers $t^{-\alpha-n}$, $n \in \mathbb{N}$. We will see many others. We call such stable scale a $\star$-asymptotic scale, and that will be defined formally in subsection 3.2.

So, assume that $e = (e_i)_{i \in I}$ is a set of functions defining our $\star$-asymptotic scale. Then there exists a matrix $\mathcal{D} = (\mathcal{D}_{i,j})_{i,j \in I}$, such that for any i in I, the derivative De_i has asymptotic expansion $\sum_{j \in I} \mathcal{D}_{i,j} e_j$; and for any positive c, there exists a matrix $\mathcal{M}_c$, such that $M_c e_i$ has asymptotic expansion $\sum_{j \in I} (\mathcal{M}_c)_{i,j} e_j$. In particular, defining the matrix

$$\mathcal{L}_G = \sum_{0 \leqslant j \leqslant m} \frac{(-1)^j}{j!} \mu_{G,j} \mathcal{D}^j,$$

we see that $L_{G,m} e_i$ has asymptotic expansion $\sum_{j \in I} (\mathcal{L}_G)_{j,i} e_j$. Since the differential operators $L_{G,m}$ are representations of the convolution semi-group, so are the matrices $\mathcal{L}_G$. This is our second representation of the convolution semigroup, which is now made up of finite dimensional matrices. Note that compared to the first representation using differential operators, we assume the existence of a $\star$-asymptotic scale in which functions can be expanded. We will see that for usual distributions this is not a restriction at all, and it is even very natural and desirable. Also, the representation $L_{F,m}$ has dimension $m+1$, while that given by $\mathcal{L}_F$ will often be of higher dimension.

Now, if $\overline{F}$ has an asymptotic expansion $\sum_{i \in I} p_{\overline{F},i} e_i$, this expansion is encoded in the vector $p_{\overline{F}} = (p_{\overline{F},i})_{i \in I}$. Then, the asymptotic expansion of $\overline{M_c F}$ is encoded in the vector $\mathcal{M}_c p_{\overline{F}}$, which is simply a product of a matrix by a vector. Since $L_{G,m}$ is a linear operator, one can also hope that $L_{G,m} \overline{F}$ has an asymptotic expansion encoded by the vector $\mathcal{L}_G p_{\overline{F}}$. This requires an extra assumption, because, in general, one cannot differentiate an asymptotic expansion termwise. Therefore, we will introduce the notion of function of class m for a given $\star$-asymptotic scale, which ensures that $\mathcal{L}_G p_{\overline{F}}$ encodes the asymptotic expansion of $L_{G,m}$. In particular, formula (1.2.3) can be written as

$$p_{\overline{M_{c_1} F \star M_{c_2} F}} = (\mathcal{L}_{M_{c_1} F} \mathcal{M}_{c_2} + \mathcal{L}_{M_{c_2} F} \mathcal{M}_{c_1}) p_{\overline{F}}.$$

This means that we can calculate the asymptotic expansion of $M_{c_1} F \star M_{c_2} F$ in a $\star$-asymptotic scale using multiplication and addition of matrices and vectors. The advantage of using this matrix representation is that we are now in the realm of linear algebra, where computers can be used both to do numerical and formal work. This is the key to efficient derivations of expansions in practical cases. One last point worth mentioning with regard to this algebraic approach is that the complexity of argument to obtain a higher-order expansion is essentially the same as that to obtain a second-order expansion. The effect of higher-order is to raise the dimension of the space one works with; but, for second-order and beyond, the spaces used will all have dimension at least 2. Only a first-order result for which a 1-dimensional space suffices can attain a significant reduction in effort. In the same vein, we note that expansions on densities and their derivatives are obtained with little extra effort. Furthermore, these results are useful, for example, in discussing the von Mises condition for infinite-order weighted averages — see subsection 4.2.

Of course, all this heuristic elaboration is rather formal at this point, and clearly the realization of this program requires us to define the proper analytical setup — this will be done in subsection 2.4 — as well as the algebraic machinery — to be done in part in subsection 2.3 — simply to be able to state the proper theorem in subsection 2.5.

Up to now we have mainly presented the algebraic formalism. It is clear that in order to justify rigorously these heuristic arguments, one has to connect this formalism to analysis. As mentioned earlier, this is mostly done in sections 5, 6, 7, 8 and 9. At this point, it is not useful to outline the proof; such outline is given in section 6.2. But we mention that the proof is by induction on the number of summands, and then a limiting argument as the number of summands goes to infinity. Of course, any limiting argument in this type of asymptotics requires a good error bound between the original functions and their approximations; on the other hand, induction somewhat requires tractable bounds. Those two requirements, sharpness and simplicity, tend to be in opposition. The key to obtain both features will be a functional analytic approach. We will write the convolution in terms of some operators, and the study of those operators will give us good bounds.

The many steps alluded to in our heuristic arguments may suggest that particularly severe restrictions will be needed on the distribution function of the X_i's and the constants c_i. However, this is not the case and the main result is rather sharp. Before going any further, we mention that our asymptotic expansions involve derivatives and moments. It will be clear from the proof that some regularity of the underlying distribution must be assumed in order to obtain even a two terms expansion. The regularity assumption has a great impact on the form of the expansion. The assumption chosen hereafter, namely differentiability up to a certain order, seems the most natural one for applications to concrete distributions. In the same vein, Barbe and McCormick (2005) shows that when α is 1 and the first moment is infinite, the truncated first moment is involved in a two terms expansion for finite convolution; moreover, the expansion takes a very different form if α is less than 1. It is quite clear from Barbe and McCormick (2005) how some results of the present paper can be formally modified when one does not have enough moments. The overall philosophy of this work is to prove that in a natural setting, limit as the number of summands tends to infinity and differentiation can be permuted with asymptotic expansions. It is however a daunting task — if not hopeless — to prove a general theorem which could cover all possible cases. In particular, it is useful to remember that even for the convolution of two heavy tail distributions, there is no known second-order formula that covers every possible case; however, we will provide some formulas or techniques which cover most distributions of interest in applications, if not all.

CHAPTER 2

Main result.

To state our main result, we need to introduce two objects: the class of distribution functions we are dealing with and some differential operators which play a role in writing the asymptotic expansions. This is the purpose of subsections 2.3 and 2.4. The main result of this paper is given in subsection 2.5. Some notation is introduced in subsection 2.1 and some facts on asymptotic scale are recalled in subsection 2.2.

2.1. Some notation.

We write $\mathbb{1}\{\cdot\}$ to denote the indicator function of a set $\{\cdot\}$.

Throughout this paper we let M_c denote the multiplication operator on distribution functions corresponding to the multiplication of random variables by c. Hence, if F is a distribution function and X is distributed according to F,

$$M_c F(t) = P\{\, cX \leqslant t \,\} = \begin{cases} F(t/c) & \text{if } c > 0, \\ \overline{F}(t/c-) & \text{if } c < 0, \\ \mathbb{1}\{\, t \geqslant 0 \,\} & \text{if } c = 0. \end{cases}$$

In the sequel, all distribution functions will be assumed ultimately continuous, thereby obviating the need for left limit notation for large values of their argument. For a general function h and positive c, we also define $M_c h$ to be $h(\cdot/c)$. Note that for c positive, $\overline{M_c F} = M_c \overline{F}$, while for c negative, $\overline{M_c F} = M_{-c}\overline{M_{-1} F} = F(\cdot/c)$.

Throughout this paper we will be interested in weighted sums of the form $\sum_{i \in \mathbb{Z}} c_i X_i$ such that $\sum_{i \in \mathbb{Z}} |c_i| E|X_i|$ is finite. By a theorem of Lévy (1954, §46), such series is commutatively convergent in distribution; that is, for any one-to-one mapping ϑ of $\mathbb{Z}$ to $\mathbb{Z}$, the random series $\sum_{i \in \mathbb{Z}} c_i X_i$ and $\sum_{i \in \mathbb{Z}} c_{\vartheta(i)} X_{\vartheta(i)}$ have the same distribution. In particular the infinite weighted convolution $\star_{i \in \mathbb{Z}} M_{c_i} F$ does not depend on the ordering of the sequence $(c_i)_{i \in \mathbb{Z}}$. This leads us to identify a sequence with its associated multiset, a terminology which we now explain.

A multiset $C = (S, \varphi)$ is a set S and a function φ, the multiplicity function, mapping S into $\mathbb{N} \cup \{\infty\}$. An element s of S belongs to the multiset if its multiplicity is at least 1, in which case we write $s \in C$. We say that the multiset is countable if it has countably many elements (even though S may be uncountable). Intuitively, a countable multiset is a sequence for which the ordering does not matter, or a countable set whose elements can be written with repetition. Note that our definition of a multiset differs slightly from the usual one in that we allow the multiplicity to be infinite or to vanish (see, for example, Aigner, 1979).

For a real valued function f defined on the elements of a multiset $C = (S, \varphi)$ and such that $\sum_{s \in S \,;\, \varphi(s) > 0} \varphi(s) |f(s)|$ is finite, we define the summation of f over C by

$$\sum_{c \in C} f(c) = \sum_{s \in S \,;\, \varphi(s) > 0} \varphi(s) f(s).$$

In particular, we say that a real multiset C is summable if $\sum_{c \in C} |c|$ is finite.

The rationale for these conventions is that we can associate a multiset to a sequence as follows. Let $(c_i)_{i \in \mathbb{Z}}$ be a sequence. Set S to be the real line $\mathbb{R}$, and, for any real number r different than 0, define $\varphi(r)$ to be the multiplicity of r in the sequence $(c_i)_{i \in \mathbb{Z}}$. Note that in order to be able to associate a multiset to a sequence, we need to allow the multiplicity of an element to be possibly infinite or to vanish. Note too that the reverse association is not unique. However, for a summable multiset $\mathcal{C}$, the number of elements in any set of the form $\mathbb{R} \setminus [-x, x]$, x positive, is finite; this fact allows one to construct an associated sequence in a natural way.

Of importance to us will be the possibility of removing an element from a multiset. If c is an element of a multiset $C = (S, \varphi)$, we define $C \setminus c$ to be the multiset whose multiplicity coincides with that of C on $S \setminus \{c\}$ and is equal to $\varphi(c) - 1$ at c. The multiset notation provides us with a convenient way to delete a single occurence of a term in a sequence by making reference only to its value rather than its position in the sequence.

The association of a multiset to a sequence and the traditional use of sequences instead of multisets in the problems we are interested in will often lead us to identify implicitly a sequence with its associated multiset. In particular, if $C = (c_i)_{i \in \mathbb{Z}}$ is a sequence, we will use the same symbol C for its associated multiset and write freely $C \setminus c_i$ or $C \setminus c$ when c is an element of the sequence or the associated multiset. The main motivation for this identification is to obtain felicitous formulas, suggestive yet synthetic.

Thanks to Lévy's (1954, §46) theorem, for a summable multiset $C = (\mathbb{R}, \varphi)$ and a distribution function F with finite mean, we can define the weighted convolution $F_C = \star_{c \in C} M_c F$ by $\star_{r \in \mathbb{R}} (M_r F)^{\star \varphi(r)}$. This is well defined because $\varphi(r)$ vanishes except at countably many real numbers and $(M_r F)^{\star 0}$ is the point mass at the origin for any real number r. In particular if the multiset C comes from a sequence $(c_i)_{i \in \mathbb{Z}}$, then F_C is $\star_{i \in \mathbb{Z}} M_{c_i} F$ as defined previously. Also if c is an element of the sequence $(c_i)_{i \in \mathbb{Z}}$, say $c = c_j$, then $F_{C \setminus c}$ or $F_{C \setminus c_j}$ is the distribution of the weighted sum $\sum_{i \in \mathbb{Z} \setminus \{j\}} c_i X_i$.

2.2. Asymptotic scales.

In this subsection, we present a few facts on terminology and notations related to asymptotic expansions. Those are needed to understand the statement of our main result and to derive asymptotic expansions efficiently. We refer to Olver (1974) for a complete account on asymptotic scales and their use.

Let N be either a positive integer or $+\infty$. A family of functions e_i, $0 \leqslant i < N$, is called an asymptotic scale if whenever $i+1$ is less than N, the asymptotic relation

$e_{i+1} = o(e_i)$ — i.e. $e_{i+1}(t) = o(e_i(t))$ as t tends to infinity — holds. We say that a function f has an $(n+1)$-terms asymptotic expansion in the scale $(e_i)_{0 \leqslant i < N}$ if there exist real numbers $p_{f,i}$, $0 \leqslant i \leqslant n$, such that

$$f(t) = \sum_{0 \leqslant i \leqslant n} p_{f,i} e_i(t) + o(e_n(t))$$

as t tends to infinity. We then write

$$f \sim \sum_{0 \leqslant i \leqslant n} p_{f,i} e_i \,.$$

When n vanishes, this notation agrees with the more usual one when one writes $f \sim g$ to mean that f/g tends to 1 at infinity. Sometimes, it will be more convenient to index an asymptotic scale by an ordered set, and an obvious variation of the definition will be understood.

It is important to note that once the asymptotic scale is chosen, the asymptotic expansion of a given function in that scale, if it exists, is unique. So, once an asymptotic scale is chosen, the map $f \mapsto p_f$ is well defined. It is a linear map, with values in $\mathbb{R}^N$, where N is the number of terms in the expansion of f. Another way to write the expansion is to think of p_f as a vector in $\mathbb{R}^N$ and the scale $e = (e_i)_{0 \leqslant i < N}$ as a vector-valued function. Then, $\sum_{0 \leqslant i < N} p_{f,i} e_i$ is the inner product of vectors $\langle p_f, e \rangle$.

For our purposes, the definition of an asymptotic expansion is not always sufficient. Following Olver (1974), we introduce the notion of generalized asymptotic expansion with respect to an asymptotic scale e_i, $0 \leqslant i < N$. If there are functions ϕ_s, $0 \leqslant s \leqslant n$, such that for any nonnegative integer k at most n,

$$f(t) = \sum_{0 \leqslant i \leqslant k} \phi_i(t) + o(e_k(t)) \,,$$

we say that $\sum_{0 \leqslant i \leqslant n} \phi_i(t)$ is a $(n+1\text{-terms})$ generalized expansion with respect to the scale e_i, $0 \leqslant i < N$, and the developing functions ϕ_i. We write

$$f \sim \sum_{0 \leqslant i \leqslant n} \phi_i \qquad (e_n) \,,$$

or simply

$$f \sim \sum_{0 \leqslant i \leqslant n} \phi_i$$

if the scale is understood. Note that there is no a priori understanding that $e_k = o(\phi_i)$ for any i between 0 and n; in theory, it could be that the only information conveyed in the expansion is that $f = o(e_k)$. Note that a generalized expansion may not be unique, even if one fixes the functions ϕ_i.

One should also be aware that the dependence of an asymptotic expansion on the asymptotic scale may be an important matter. For instance, write $\overline{\Phi}$ for the tail of the standard normal distribution and consider the asymptotic scale $e_i = \overline{\Phi}^i$,

$i \in \mathbb{N}^*$. Define the function $f(t) = \overline{\Phi}(t) + \overline{\Phi}^2(t)$. One has the asymptotic expansion

$$f \sim e_1 + e_2.$$

Clearly, this approximation cannot be improved. On the other hand, if one decides to use the asymptotic scale $\tilde{e}_i(t) = \overline{\Phi}(t)/t^i$, $i = 0, 1, 2, \ldots$, one has for any nonnegative n

$$f \sim \tilde{e}_0 \qquad (\tilde{e}_n).$$

In this asymptotic expansion the part of f coming from e_2 is not taken into consideration, despite the fact that in the chosen asymptotic scale, the expansion has as many terms as one wishes. Finally, if one decides to use the asymptotic scale $\overline{e}_i(t) = e^{-t^2/2}/t^i\sqrt{2\pi}$, $i \geqslant 1$, then

$$f \sim \overline{e}_1 + \sum_{i \geqslant 1} (-1)^i \frac{(2i-1)!}{2^{i-1}(i-1)!} \overline{e}_{2i+1}.$$

Again, in the asymptotic scale $\overline{e}_i$, $i \geqslant 1$, this asymptotic expansion has as many terms as one wishes though the term in e_2 in f is not taken into consideration. The term e_2 is effectively 0 at every level of scale with respect to the asymptotic scale $(\tilde{e}_i)_{i \geqslant 0}$ or the asymptotic scale $(\overline{e}_i)_{i \geqslant 1}$. Moreover, this last asymptotic expansion is a divergent series, which gives a useless approximation if one does not truncate it.

We highly recommend the short first chapter of Olver's (1974) book, where many simple examples are discussed; very useful, not so useful and totally useless expansions are shown; and the pros and cons of asymptotic expansions are well presented.

2.3. The Laplace characters.

Let D^k be the differential operator defined by $D^k h(x) = h^{(k)}(x)$ for any k times differentiable function h. As usual, we set D^0 to be the identity and $D^1 = D$.

Recall that a linear differential operator with constant coefficients is a finite sum $\sum_{0 \leqslant i \leqslant m} p_i D^i$. This is a polynomial in D, and the order of the operator is the degree of the polynomial. Thus, p_m is nonzero if and only if the previous differential operator is of order m. We write $\mathbb{R}_m[D]$ for the set of all linear differential operators with constant coefficients and order at most m. It is naturally endowed with a vector space structure, corresponding to that of the polynomials. We will be mostly interested in a ring structure though. In $\mathbb{R}_m[D]$, we define a composition as follows. If

$$p = \sum_{0 \leqslant i \leqslant m} p_i D^i \qquad \text{and} \qquad q = \sum_{0 \leqslant i \leqslant m} q_i D^i,$$

are in $\mathbb{R}_m[D]$, we set

$$pq = \sum_{0 \leqslant i \leqslant m} \Big(\sum_{0 \leqslant j \leqslant i} p_j q_{i-j} \Big) D^i.$$

In other words, we use for the ring structure of $\mathbb{R}_m[D]$ that of the quotient ring of the polynomials in D modulo the ideal generated by D^{m+1}.

In the following, we use the notation $\mu_{F,k}$ for the moment of order k of a distribution function F, that is $\int x^k \, \mathrm{d}F(x)$. In particular, $\mu_{F,0}$ is 1.

DEFINITION. *Let F be a distribution function and m be an integer such that F has a finite m-th absolute moment. The m-th Laplace character of F is the element of $\mathbb{R}_m[\mathrm{D}]$ defined by*

$$L_{F,m} = \sum_{0 \leqslant k \leqslant m} \frac{(-1)^k}{k!} \mu_{F,k} \, \mathrm{D}^k \, .$$

When m is clear from the context, we may simply write L_F. Observe that L_{δ_0} is the identity. The origin of the term 'Laplace character' will be explained after our next proposition.

REMARK. Laplace characters are defined as elements in the ring $\mathbb{R}_m[\mathrm{D}]$. In particular, except specified otherwise, operations on the Laplace characters are those in that quotient ring. In formulas involving Laplace characters and functions, the operations on Laplace characters have the highest precedence, meaning for instance that $L_{F,m} L_{G,m} h$ is $(L_{F,m} L_{G,m}) h$.

PROPOSITION 2.3.1. *If F and G are two distribution functions with finite absolute moment of order m, then $L_{F,m} L_{G,m} = L_{F \star G, m}$.*

PROOF. The binomial formula shows that

$$\sum_{0 \leqslant i \leqslant k} \frac{\mu_{F,i}}{i!} \frac{\mu_{G,k-i}}{(k-i)!} = \frac{\mu_{F \star G, k}}{k!} \, .$$

The conclusion follows in an easy way from the definition of the product. ∎

The previous proposition asserts that the Laplace characters are morphisms of semigroups. Since the maps $F \mapsto \mu_{F,k}$ are linear, the Laplace characters are representations of convolution algebras in $\mathbb{R}_m[\mathrm{D}]$. One may wonder if other representations exist, that depend on the distribution through some other set of characteristics, say for instance, moments of fractional order or quantiles. In fact, the answer is negative. By a result of Mattner (2004), the mapping of a distribution with m moments to its first m cumulants is universal among all continuous homomorphisms of the convolution algebra of distributions with m moments into Hausdorff topological groups. This implies in particular that one can express the Laplace characters in terms of cumulants.

We can now explain our choice of the name 'Laplace character'. Formally, $L_{F,m}$ is the m-th Taylor polynomial of the Laplace transform of F where we substitute the operator D for the variable. Hence the 'Laplace' part of the name. Equivalently, we could define $L_{K,m}$ to be $E e^{-Y\mathrm{D}}$ in $\mathbb{R}_m[\mathrm{D}]$, where Y has distribution function K — note that with respect to Mattner's (2004) result, it is clear from this definition that the Laplace characters can be expressed in terms of cumulants. Moreover, Proposition 2.3.1 shows the Laplace character obeys a property very similar to the Laplace transform, namely that this trivializes the convolution into a product of polynomials (modulo an ideal here). This property is very similar to that defining the characters of a group, hence the name.

One needs to be careful with one subtlety, namely that our multiplication in $\mathbb{R}_m[\mathrm{D}]$ is not the usual composition of operators acting on functions. If T_1 and T_2 are two operators, we write, in this paragraph only, $T_2 \circ T_1$ for their usual composition, that is $(T_2 \circ T_1)h = T_2(T_1 h)$ for any smooth function h. We then see that if $i + j > m$, we have $\mathrm{D}^i \circ \mathrm{D}^j = \mathrm{D}^{i+j}$, while, $\mathrm{D}^i \mathrm{D}^j = 0$ in $\mathbb{R}_m[\mathrm{D}]$. However, the usual composition $\mathrm{D}^i \circ \mathrm{D}^j$ is the product $\mathrm{D}^i \mathrm{D}^j$ modulo an operator in the ideal generated by D^{m+1}. This is in fact important (see subsection 6.2). We could avoid this subtlety by defining operators acting on an infinite sequence of tuples of the form $(\overline{F}_i, \overline{F}_i^{(1)}, \ldots, \overline{F}_i^{(m)})$, $i \geqslant 1$. It is really a matter of taste, and we feel that the framework proposed here is somewhat more convenient for our purposes.

For computational purposes, it is of interest to note that operators in $\mathbb{R}_m[\mathrm{D}]$ which are not in the ideal generated by D — in other words, those which, as a polynomial in D, have a nonzero constant term — can be inverted. That is to say, $\big(\mathbb{R}_m[\mathrm{D}], (\mathrm{D})\big)$ is a local ring. In particular, the Laplace characters can be inverted. Two explicit inversion formulas are given in section 5.1.

It is also clear that there are other ways of thinking of the Laplace characters and their inverses that can be useful in applications. For instance we can think of $L_{F,m}$ as the m-th Taylor polynomial of the Laplace transform L_F (if it exists) of F evaluated at D. Then $L_{F,m}^{-1}$ is simply obtained by taking the m-th Taylor polynomial of $1/L_F$ and evaluating it in the variable D.

There are other algebraic operations on Laplace characters which are of interest. For instance, we will use the Mellin-Stieltjes convolution between two distribution functions F and G on the nonnegative half-line. It is written $F \stackrel{\mathrm{M}}{\star} G$, and defined by

$$F \stackrel{\mathrm{M}}{\star} G(t) = \int_0^\infty F(t/x) \, \mathrm{d}G(x) \, .$$

If X and Y are two independent random variables with respective distributions F and G, their product XY has distribution $F \stackrel{\mathrm{M}}{\star} G$. Since $E(XY)^j = EX^j EY^j$, the Laplace character of $F \stackrel{\mathrm{M}}{\star} G$ is obtained by multiplying the Laplace characters of F and G coefficient-wise, with alternating signs; that is, defining the operation $\stackrel{\mathrm{M}}{\circ}$ in $\mathbb{R}_m[\mathrm{D}]$ by

$$\mathrm{D}^i \stackrel{\mathrm{M}}{\circ} \mathrm{D}^j = \begin{cases} (-1)^i \mathrm{D}^i & \text{if } i = j, \\ 0 & \text{otherwise}, \end{cases}$$

and extending this operation by bilinearity,

$$L_{F \stackrel{\mathrm{M}}{\star} G, m} = \sum_{0 \leqslant j \leqslant m} \frac{(-1)^j}{j!} \mu_{F,j} \mu_{G,j} \mathrm{D}^j = L_{F,m} \stackrel{\mathrm{M}}{\circ} L_{G,m} \, .$$

Yet, another algebraic operation that one could consider, and which we will not use in this paper, is that of differentiating formally the Laplace characters as polynomials in D. Assuming that $\mu_{F,k}$ exists and does not vanish, writing

$dG(x) = \mu_{F,k}^{-1} x^k\, dF(x)$, we see that $L_{G,m}$ is $(-1)^k/\mu_{F,k}$ times the k-th derivative with respect to the variable D of $L_{F,m+k}$.

Similarly, if F is a distribution on the nonnegative half-line and its normalized integrated distribution is $G(t) = \mu_{F,1}^{-1} \int_0^t \overline{F}(x)\, dx$, then

$$L_{G,m} = -\mu_{F,1}^{-1}(L_{F,m+1} - \mathrm{Id})/\mathrm{D}\,.$$

The interest of such algebraic formulas is to produce a form of operational calculus to derive tail expansions. This will be clear in section 3.

2.4. Smoothly varying functions of finite order.

For any real number x and any positive integer k, we write $(x)_k$ for the falling factorial $x(x-1)\cdots(x-k+1)$. We also set $(x)_0 = 1$. If h is a function, $h^{(k)}$ is the k-th derivative of h if it exists, with the usual convention $h^{(0)} = h$.

Recall that a function h defined in some neighborhood of infinity is smoothly varying with negative index $-\alpha$ if it is ultimately infinitely differentiable and

$$\lim_{t\to\infty} t^k h^{(k)}(t)/h(t) = (-\alpha)_k \tag{2.4.1}$$

for every integer k (see Bingham, Goldie and Teugels, 1989, §1.8). The set of all smoothly varying functions with index $-\alpha$ is written $SR_{-\alpha}$. The set of smoothly varying functions of a given finite order which we will define contains $SR_{-\alpha}$ and can be thought of as a Sobolev space in the framework of regular variation. Before proceeding, it is useful to remark that relation (2.4.1) forces $h^{(k)}$ to be regularly varying of index $-\alpha-k$ (see Bingham, Goldie and Teugels, 1989, Proposition 1.8.1).

DEFINITION. *Let m be a positive integer. A function h is smoothly varying of index $-\alpha$ and order m if it is ultimately m-times continuously differentiable and $h^{(m)}$ is regularly varying of index $-\alpha-m$. We write $SR_{-\alpha,m}$ for the set of all such functions.*

From Karamata's theorem, we see that a function h in $SR_{-\alpha,m}$ satisfies (2.4.1) for any $0 \leqslant k \leqslant m$. Note that if m_1 is at most m_2 then $SR_{-\alpha,m_2}$ is included in $SR_{-\alpha,m_1}$.

To define $SR_{-\alpha,s}$ for a real number s, we write δ_x for the point mass at x. Thus if h is a function, $\delta_x h = \int h\, d\delta_x = h(x)$. We then define the operator

$$\Delta_{t,x}^r = \mathrm{sign}(x) \frac{\delta_{t(1-x)} - \delta_t}{|x|^r \delta_t}\,.$$

In other words, for any function g we set

$$\Delta_{t,x}^r g = \mathrm{sign}(x) \frac{g\bigl(t(1-x)\bigr) - g(t)}{|x|^r g(t)}\,.$$

DEFINITION. *Let s be a positive real number. Write $s = m + r$ where m is the integer part of s and r is in $[0, 1)$. A function h is smoothly varying of index $-\alpha$ and order s if it belongs to $SR_{-\alpha,m}$ and*

$$\lim_{\delta \to 0} \limsup_{t \to \infty} \sup_{0 < |x| \leqslant \delta} |\Delta_{t,x}^r h^{(m)}| = 0. \tag{2.4.2}$$

Note that $m = m + 0$, and that if h belongs to $SR_{-\alpha,m}$ then (2.4.2) holds for $r = 0$ thanks to the uniform convergence theorem on compact subsets of $(0, \infty)$ (Bingham, Goldie, Teugels, 1989, Theorem 1.2.1).

The following result provides a simple sufficient condition for a function to be smoothly varying of a given order.

PROPOSITION 2.4.1. *Let s be a positive real number, and let m be its integer part. A sufficient condition for a function h to be smoothly varying of index $-\alpha$ and order s is that $h^{(m+1)}$ is regularly varying, of index $-\alpha - m - 1$.*

PROOF. The result follows from Proposition 5.2.1. ∎

In particular, Proposition 2.4.1 shows that all distributions with regularly varying tails used in applications have a smoothly varying tail of arbitrary order.

When s is smaller than 1, the set $SR_{-\alpha,s}$ is closely related to that of all the functions satisfying the Lipschitz condition $[\mathrm{D}_s]$ of Borovkov and Borovkov (2002), while for positive m it is related to their condition $[\mathrm{D}_m]$.

Smoothly varying functions of finite order are studied further in section 5.2.

2.5. Asymptotic expansion for infinite convolution.

To state our main result, recall that the ℓ_p norm of a sequence or a multiset C is

$$\|C\|_p = \left(\sum_{c \in C} |c|^p \right)^{1/p}.$$

This defines a norm only if p is at least 1. Nevertheless, we will still use the same notation with the same meaning when p is less than 1. When p is infinite, $\|C\|_\infty$ is defined to be the supremum of the elements of the multiset. Given three nonnegative numbers α, γ and ω, we define

$$N_{\alpha,\gamma,\omega}(C) = \|C\|_{\gamma(\frac{\alpha}{\alpha+\omega} \wedge \frac{1}{2})} \vee 2^{\alpha/(\alpha+\omega)} \|C\|_\infty.$$

This may or may not be a norm, according to whether or not $\gamma\alpha/(\alpha + \omega)$ and $\gamma/2$ are at least 1. For the values of γ that we will use, that is γ positive and less than 1, this is not a norm.

The next theorem is the main result of the paper. It establishes a generalized asymptotic expansion for some infinite weighted convolutions with respect to the asymptotic scale $\mathrm{Id}^{-i}\overline{F}$, $i \geqslant 0$. To state this result, we need further notation.

Note that if c is negative, then the upper tail of cX_i is driven by the lower tail of X_i. This induces some complications in stating a result for the tail of weighted convolutions because one needs stronger assumptions when the signs of the constants and the random variable can be arbitrary. This explains the formulation of the next result. It does not cover all the possible variations, but seems to give what is needed in most applications. Other cases often may be obtained by simple changes in the proof.

For two functions f and g, we write $f \asymp g$ to mean that the ratio f/g is ultimately bounded away from 0 and infinity.

Recall that for a multiset C and a distribution function F, we write F_C for the infinite weighted convolution $\star_{c \in C} M_c F$, assuming that this infinite convolution product converges commutatively to a nondegenerate limiting distribution. Recall also that for an element c of the multiset, $C \backslash c$ is the multiset obtained by 'removing' one copy of c, that is decreasing by 1 the counting function of C at c.

THEOREM 2.5.1. *Let ω be at least 1. Let F be a distribution function with $\overline{F}$ in $SR_{-\alpha,\omega}$ and $\overline{M_{-1}F} = O(\overline{F})$ at infinity. Let m and k be two integers such that m is smaller than α, and $m+k$ is smaller than ω. Furthermore, if we are to consider weighted convolution with possibly negative weights, assume either that $\overline{M_{-1}F}$ is also in $SR_{-\alpha,\omega}$ and $\overline{M_{-1}F} \asymp \overline{F}$ or that F vanishes in a neighborhood of $-\infty$.*

Let γ be a positive number less than $\omega - m - k$ and 1. Then, there exists a function $\eta(\cdot)$ converging to 0 at infinity and a real number t_0 such that for any t at least t_0, for any multiset C with $N_{\alpha,\gamma,\omega}(C) \leqslant 1$,

$$\left| \overline{F}_C^{(k)} - \sum_{c \in C} L_{F_{C \backslash c}, m} \overline{M_c F}^{(k)} \right| \leqslant \mathrm{Id}^{-m-k} \overline{F} \eta$$

on $[t_0, \infty)$. In particular for any multiset C with $N_{\alpha,\gamma,\omega}(C)$ finite,

$$\overline{F}_C^{(k)} \sim \sum_{c \in C} L_{F_{C \backslash c}, m} \overline{M_c F}^{(k)} \qquad (\mathrm{Id}^{-m-k} \overline{F}).$$

The Laplace characters are polynomials in D. A monomial D^j applied to $\overline{M_c F}^{(k)}$ yields a function of order $\overline{F}^{(j+k)}$, that is of order $\overline{F}/\mathrm{Id}^{j+k}$. Consequently, despite its appearance, the conclusion of Theorem 2.5.1 is an explicit expansion; this point will be clarified in subsection 3.1. The remaining issues are how to calculate the infinite series in practice, and, assuming that $\overline{F}$ has an expansion in an asymptotic scale, to determine if $\overline{F}_C^{(k)}$ has an asymptotic expansion in the same scale, and calculate that expansion from that of $\overline{F}$. These issues will be dealt with in section 3.

Theorem 2.5.1 calls for several remarks.

REMARK. By allowing a certain uniformity in the Potter type bounds and in the asymptotic smoothness assumption, a degree of uniformity in our result with respect to the underlying distribution F can be achieved. However, we choose not to develop such a refinement since it would entail a greater level of technicality which may distract from the main aim of the paper.

REMARK. When some elements of the multiset are negative, the assumption $\overline{F} \asymp \overline{M_{-1}F}$ is not necessary. It is quite clear from the proof how to modify the statement if we assume that $\overline{M_{-1}F}$ is in some $SR_{-\beta,\omega'}$. However, the statement is far nicer if either both tails are comparable, or if F vanishes in a neighborhood of $-\infty$.

REMARK. When the c's are nonnegative, one does not really need to assume $\overline{M_{-1}F} = O(\overline{F})$. Certainly, the result holds provided only that the m-th absolute moment of the distribution is finite. This technical point appears in Lemma 7.3.1.

REMARK. If either k or m are positive and their sum is less than ω, then ω is more than 1, and the first sentence in Theorem 2.5.1 is not needed. Assuming that ω is at least 1 is only needed when both k and m vanish; it is assumed only to ensure that $\overline{F}$ and possibly $\overline{M_{-1}F}$ are normalized regularly varying.

REMARK. The analoguous statement on the lower tail holds as well.

REMARK. Since the Laplace characters are differential operators with constant coefficients, they commute with the derivative D. Consequently, Theorem 2.5.1 implies that for a certain class of distributions and sequences, taking limit with respect to the number of nonvanishing terms in the sequence C and differentiation can be permuted in an asymptotic expansion. It is quite striking that this can be done with some uniformity with respect to a rather large class of sequences.

REMARK. A suggestive way of writing the expansion of Theorem 2.5.1 is by introducing fractions very much inspired by the algebraic construction of modules of fractions as in Atiyah and MacDonald (1969). The construction is as follows. Whenever k is a nonnegative integer less than m, the differentiation D maps $SR_{-\alpha-k,m-k}$ into $SR_{-\alpha-k-1,m-k-1}$. In this remark only, we define D on $SR_{-\alpha-m,0}$ as the constant operator mapping any function to 0. In a compact notation

$$ SR_{-\alpha,m} \xrightarrow{\text{D}} SR_{-\alpha-1,m-1} \xrightarrow{\text{D}} \cdots \xrightarrow{\text{D}} SR_{-\alpha-m,0} \xrightarrow{\text{D}} 0. $$

The Laplace characters are closed under multiplication. Given a function f in $\cup_{0 \leqslant k \leqslant m} SR_{-\alpha-k,m-k}$ and a Laplace character $L_{F,m}$, consider the formal fraction $f/L_{F,m}$. These fractions are added according to the rule

$$ \frac{f}{L_{F,m}} + \frac{g}{L_{G,m}} = \frac{L_{G,m}f + L_{F,m}g}{L_{F,m}L_{G,m}}. $$

Write num( ) for the numerator of a fraction, that is $\text{num}(f/L_{F,m}) = f$. Theorem 2.5.1 implies that

$$ \overline{F}_C^{(k)} = \text{num}\Big(\sum_{c \in C} \frac{\overline{M_c F}^{(k)}}{L_{M_c F, m}} \Big) + o(\text{Id}^{-m-k}\overline{F}). $$

REMARK. It will be clear in the sequel that the formulation of Theorem 2.5.1 is well suited for applications. One may still wonder if one can relax its assumptions

2.5. ASYMPTOTIC EXPANSION FOR INFINITE CONVOLUTION

on F. The example of first-order autoregressive processes suggests that some refinements may be given, but not in a fundamental way. Indeed, consider the sequence $C = (c_i)_{i \in \mathbb{Z}}$ defined by $c_i = a^i$ for nonnegative integers i, and $c_i = 0$ for negative i. Thus, F_C is the distribution of $Y = \sum_{i \geqslant 0} a^i X_i$. For any integer k, set

$$Z_{k,i} = \sum_{0 \leqslant j < k} a^j X_{ki+j}\,.$$

For fixed k, as i runs through the nonnegative integers, these random variables are independent and equidistributed. Moreover, $Y = \sum_{i \geqslant 0} a^{ki} Z_{k,i}$. Hence, if the distribution function of $Z_{k,i}$ fulfills the assumptions of Theorem 2.5.1, one can obtain a tail expansion of F_C, even though the distribution of X_i may not satisfy the assumptions of Theorem 2.5.1. This may happen for instance if the distribution of X_i is a mixture containing point masses. Note however that the tail expansion will be expressed in terms of the distribution of $Z_{k,i}$ and not of X_i. One would then need to relate the derivative of the distribution function of $Z_{k,i}$ to that of X_i. For the problem at hand, if $Z_{k,i}$ has a density for k large enough, then by increasing k, one can assume that this density is smooth. However, Y could still have a density, even if, regardless how large k is, $Z_{k,i}$ does not have one. As mentioned in the introduction, Solomyak's (1995) work is a vivid reminder on how difficult it is to study the marginal distribution of such a basic time series model.

For more general linear processes, one could imagine using similar blocking techniques. One would then need to extend Theorem 2.5.1 by replacing the $M_c F$'s by some more general F_i's, each F_i being in $SR_{-\alpha,\omega}$. Formally, the expression is obvious to obtain. A proof along the line of Theorem 2.5.1 is possible under some assumptions on the F_i's. However, it is not clear that such a refinement presents much interest in applications. There is in fact a great similarity with the central limit theorem for densities, where one does not need the summands to have a density, but the characteristic function to be in some L^r space with some r at least 1 (see Feller, 1971, §XV.5). Yet, in most real life applications — if not all — the local central limit theorem is applied when the summands have a well behaved density. The same seems true about the conditions of Theorem 2.5.1.

CHAPTER 3

Implementing the expansion.

In this section we illustrate Theorem 2.5.1 and its implementation through examples. In the first subsection we discuss the number of terms in the expansion given by Theorem 2.5.1. In the second subsection, we introduce a special type of asymptotic scale and a class of functions well behaved with respect to these scales. This will be used, in the third subsection, to explain how for standard statistical distributions, matrix identification of the differential and multiplication operators provides a computationally efficient way to derive tail expansions. In the fourth subsection, our construction is illustrated on some examples. In the fifth subsection, we discuss an open problem related to a second-order formula for weighted convolutions under the assumption of regular variation with remainder. The last subsection contains some open problems.

In what follows, if the multiset C is nonnegative, meaning all its elements are nonnegative, we write C_p for $\sum_{c \in C} c^p$. When C_{p+q} as well as C_p and C_q are finite, we write $C_{p;q}$ for $C_{p+q} - C_p C_q$.

3.1. How many terms are in the expansion?

Let us consider the expansion given in Theorem 2.5.1 for the tail of the distribution function, that is when k vanishes. It asserts

$$\overline{F_C} = \sum_{c \in C} L_{F_{C \setminus c}, m} \overline{M_c F} + o\bigl(\mathrm{Id}^{-m} \overline{F}\bigr). \tag{3.1.1}$$

It seems that this formula provides an $(m+1)$-term expansion. However, one should remember that the number of terms in an expansion depends crucially on the asymptotic scale chosen. For the case at hand, we will show that, for natural scales, this formula may give more or less than m terms. We suspect that this fact explains the failure of previous purely analytical attempts to find only a two terms expansion in the general infinite-order case. The algebraic flavor of the Laplace characters will be developed further in the next subsection, and will be the key to efficiently deriving asymptotic expansions of specific distributions.

When m is 1, formula (3.1.1) is

$$\overline{F_C} = \sum_{c \in C} (\overline{M_c F} - \mu_{F_{C \setminus c}, 1} \overline{M_c F'}) + o\bigl(\mathrm{Id}^{-1} \overline{F}\bigr). \tag{3.1.2}$$

For simplicity of the discussion, assume that the c's are nonnegative, and let us ask: how many terms does (3.1.2) provide?

A naive answer is 2, for the first term is $\sum_{c\in C} \overline{M_c F}$, while the second one is $-\sum_{c\in C} \mu_{F_{C\setminus c},1} \overline{M_c F'}$. The following four examples show that the truth is more complex.

Example 1. Take $\overline{F}(t) = t^{-\alpha}(1 + 1/\log t)$ for t large enough. Then for c positive,

$$\overline{M_c F}(t) = \frac{c^\alpha}{t^\alpha}\left(1 + \frac{1}{\log t - \log c}\right)$$
$$= \frac{c^\alpha}{t^\alpha} + \frac{c^\alpha}{t^\alpha \log t} + \frac{c^\alpha}{t^\alpha \log t} \frac{\log c}{\log t - \log c}.$$

Moreover, $\overline{M_c F'}(t) \sim -\alpha c^\alpha / t^{\alpha+1}$ as t tends to infinity. Applying formula (3.1.2), we obtain

$$\overline{F}_C(t) = \frac{C_\alpha}{t^\alpha} + \frac{C_\alpha}{t^\alpha \log t}(1 + o(1)).$$

The interesting point is that this two terms expansion is given by the sum $\sum_{c\in C} \overline{M_c F}$; it does not involve the term $\sum_{c\in C} \mu_{F_{C\setminus c},1} \overline{M_c F'}$.

Example 2. Take $\overline{F}(t) = t^{-\alpha} + t^{-\alpha-2}$ for t large enough, and assume that F is supported inside the nonnegative half-line. This last assumption ensures that $\mu_{F,1}$ does not vanish. The derivative of $\overline{F}$ is

$$\overline{F}'(t) = -\alpha t^{-\alpha-1} - (\alpha+2)t^{-\alpha-3}.$$

Consequently, (3.1.2) yields

$$\overline{F}_C(t) = C_\alpha t^{-\alpha} - \alpha t^{-\alpha-1} \sum_{c\in C}(C_1 - c)c^\alpha \mu_{F,1} + o(t^{-\alpha-1})$$
$$= C_\alpha t^{-\alpha} + \alpha C_{\alpha;1} \mu_{F,1} t^{-\alpha-1} + o(t^{-\alpha-1}).$$

Therefore, if $C_1 C_\alpha \neq C_{\alpha+1}$, the first term is given by a contribution from $\sum_{c\in C} \overline{M_c F}$, while the second one comes from $\sum_{c\in C} \overline{M_c F'}$.

Example 3. Take $\overline{F}(t) = t^{-\alpha} - t^{-\alpha-3}$ for large t, and assume now that F is symmetric. Hence $\mu_{F,1}$ vanishes. Formula (3.1.2) yields

$$\overline{F}_C(t) = C_\alpha t^{-\alpha} + o(t^{-\alpha-1}).$$

In some sense, the formula fails to give a two terms expansion. So we consider formula (3.1.1) when m is 2. Since $\mu_{F,1}$ vanishes, it yields

$$\overline{F}_C(t) = \sum_{c\in C} \overline{M_c F}(t) + \frac{1}{2}\sum_{c\in C} \mu_{F_{C\setminus c},2} \overline{M_c F}^{(2)}(t) + o(t^{-\alpha-2}). \tag{3.1.3}$$

Also, since $\mu_{F,1}$ is zero,

$$\mu_{F_{C\setminus c},2} = (C_2 - c^2)\mu_{F,2}.$$

Consequently,

$$\overline{F}_C(t) = C_\alpha t^{-\alpha} + \frac{1}{2}\sum_{c\in C}(C_2 - c^2)\mu_{F,2}\alpha(\alpha+1)c^\alpha t^{-\alpha-2} + o(t^{-\alpha-2})$$

$$= C_\alpha t^{-\alpha} - \frac{\alpha(\alpha+1)}{2}C_{\alpha;2}\mu_{F,2}t^{-\alpha-2} + o(t^{-\alpha-2}).$$

We see that the second term in this formula comes in fact from the third term in (3.1.1). Notice that the second term in the expansion for $\overline{F}$ played no role.

Example 4. Let us modify example 2 by writing $\overline{F}(t) = t^{-\alpha} + at^{-\alpha-2}$ for large t. And let us take F to be symmetric. Formula (3.1.1) with m equal 2 yields

$$\overline{F}_C(t) = C_\alpha t^{-\alpha} + aC_{\alpha+2}t^{-\alpha-2} + \frac{\alpha(\alpha+1)}{2}\mu_{F,2}\sum_{c\in C}(C_2 - c^2)c^\alpha t^{-\alpha-2} + o(t^{-\alpha-2})$$

$$= C_\alpha t^{-\alpha} + t^{-\alpha-2}\left(aC_{\alpha+2} - \frac{\alpha(\alpha+1)}{2}\mu_{F,2}C_{\alpha;2}\right) + o(t^{-\alpha-2}).$$

For any summable sequence C of nonnegative numbers, we can find a such that

$$aC_{\alpha+2} - \frac{\alpha(\alpha+1)}{2}\mu_{F,2}C_{\alpha;2} = 0.$$

In this case, we need to include at least one more term, taking at least $m = 3$ in formula (3.1.1) if we want two nonzero terms.

A first point of these examples is that obtaining simply a two terms expansion formula is rather hopeless in general. Clearly, some tricky cancellation may occur, depending on the particular sequence C considered as well as the coefficients in the expansion of $\overline{F}$.

A second point is that when some cancellations occur, we may need to add terms; at least if we can, since the condition m less than α caps the number of terms we can obtain. In our examples, this was easy, but in other cases, this may involve far more complicated calculations. Hence, we need more effective ways to implement the expansion, and this will be discussed in the next subsections.

A third point is that it raises an interesting question: Is it possible that for any fixed m, as large as we want, can we find a distribution for which formula (3.1.1) provides only a 1 term expansion? We address this question in the rest of this subsection. The answer is yes, though it is not a generic case. To make this clear, let us examine the following statement.

STATEMENT. *Generically, (3.1.1) with m equal 2 provides at least a two terms expansion, even if $\mu_{F,1}$ vanishes.*

When m is 2, the right hand side in (3.1.1), up to the $o(\cdot)$-term, is

$$\sum_{c\in C}\overline{M_c F} - \sum_{c\in C}\mu_{F_{C\setminus c},1}\overline{M_c F'} + \frac{1}{2}\sum_{c\in C}\mu_{F_{C\setminus c},2}\overline{M_c F''}.$$

Let σ_F^2 be the variance of F. Since

$$\mu_{F_{C\setminus c},1} = (C_1 - c)\mu_{F,1}$$
$$\mu_{F_{C\setminus c},2} = (C_2 - c^2)\sigma_F^2 + (C_1 - c)^2\mu_{F,1}^2$$

and $\overline{M_c F''}(t) \sim c^\alpha \overline{F''}(t)$, as t tends to infinity, the formula reads

$$\sum_{c \in C} \overline{M_c F} - \sum_{c \in C} (C_1 - c)\mu_{F,1} \overline{M_c F'}$$
$$- \frac{1}{2}\Big(C_{\alpha;2}\sigma_F^2 + (C_1 C_{\alpha;1} - C_{\alpha+1;1})\mu_{F,1}^2\Big)\overline{F''}(t) + o\big(\overline{F}(t)/t^2\big).$$

Generically, there is no cancellation, and even if $\mu_{F,1}$ vanishes, the term in $\overline{F''}(t)$ yields a nonzero term which (generically) does not cancel with $\sum_{c \in C} \overline{M_c F}$. So, the formula yields at least two terms.

The companion of the positive statement is as follows.

STATEMENT. *Consider the formula (3.1.1) for both m and the sequence C fixed. Then there exists a distribution (depending on m and the sequence C) for which the formula yields only a 1-term expansion.*

We will prove this statement in section 3.4, once we have an effective way of calculating expansions.

3.2. ⋆-Asymptotic scales and functions of class m.

The few examples that we gave in subsection 3.1 were calculated easily by hand. In more complicated cases as well as for theoretical reasons (for instance to prove the statement at the end of subsection 3.1), we need more effective ways to derive expansions. In Barbe and McCormick (2005), we introduced an algebraic technique, a tail calculus, which allows one to derive asymptotic expansions for finite weighted convolutions of distribution functions such that $\overline{F}(t) \sim t^{-\alpha} P(1/t)$ for some polynomial P. An extension to infinite-order moving averages was carried out for this class in Barbe and McCormick (2004). Distributions for which this calculus is applicable include the Pareto, Student and the Generalized Pareto. However, it does not apply to other well used distributions in the heavy tail literature. So we will develop a tail calculus for those. A key ingredient will be the notions of ⋆-asymptotic scale and function of class m for a ⋆-asymptotic scale, which we are going to define.

The expansion in Theorem 2.5.1 involves multiplication by constants. Both the derivative D and the multiplication operators M_c act componentwise on the scale e by $\mathrm{D}e = (\mathrm{D}e_i)_{i \in I}$ and $M_c e = (M_c e_i)_{i \in I}$. For every t, we can take linear combinations of the components of the vector $e(t)$ by left multiplying it by a matrix. For vector valued functions, we extend the expansion notation $\sim$ as acting componentwise. So, a vector valued function $f = (f_1, \ldots, f_n)$ has asymptotic expansion $g = (g_1, \ldots, g_n)$ if $f_i \sim g_i$ for each $i = 1, \ldots, n$. For a matrix A we write A^{t} for its transpose.

DEFINITION. *An asymptotic scale $e = (e_i)_{i \in I}$ is a $\star$-asymptotic scale if there exist lower-triangular matrices $\mathcal{D}$ and $\mathcal{M}_c$ such that $\mathrm{D}e \sim \mathcal{D}^{\mathrm{t}}e$ and for any c positive, $M_c e \sim \mathcal{M}_c^{\mathrm{t}} e$.*

In particular, in an asymptotic sense, a $\star$-asymptotic scale is closed under differentiation and multiplication of the variable by a positive constant.

In the definition of a $\star$-asymptotic scale, requiring that the matrices $\mathcal{D}$ and $\mathcal{M}_c$ are lower-triangular expresses the assumptions that for any function e_i of a $\star$-asymptotic scale, $\mathrm{D}e_i = O(e_i)$ and $M_c e_i = O(e_i)$ for any c. In practice, note that the assumption on the derivative holds as soon as e_i is smoothly varying of negative index and order at least 1; the assumption on the action of the multiplication operator holds whenever e_i is regularly varying. Hence, we see that for the type of problem we are dealing with, these two assumptions are easy to verify. Perhaps the simplest example of a $\star$-asymptotic scale is that defined by the functions $e_i = \mathrm{Id}^{-\alpha - i}$, $i \in \mathbb{N}$.

Our next result shows how to calculate an asymptotic expansion of a function $M_c f$ if one knows the asymptotic expansion of f in a $\star$-asymptotic scale.

PROPOSITION 3.2.1. *If f is a function having an asymptotic expansion in a $\star$-asymptotic scale, then the following diagram commutes.*

$$\begin{array}{ccc} f & \xrightarrow{M_c} & M_c f \\ \downarrow{p} & & \downarrow{p} \\ p_f & \xrightarrow{\mathcal{M}_c} & p_{M_c f} \end{array}$$

PROOF. Let $e = (e_i)_{0 \leqslant i < N}$ be a $\star$-asymptotic scale. Let n be a nonnegative integer less than N. Since f has an expansion in that scale and $M_c e_n = O(e_n)$

$$f(t/c) = \sum_{0 \leqslant i \leqslant n} p_{f,i} e_i(t/c) + o\big(e_n(t/c)\big)$$
$$= \langle p_f, \mathcal{M}_c^{\mathrm{t}} e \rangle(t) + o\big(e_n(t)\big)$$

The result follows by definition of the transpose. ∎

As pointed out in Olver (1974), in general, asymptotic expansions can be integrated but not differentiated. However, for all distributions used in applications, the asymptotic expansions of derivatives can be obtained by a termwise differentiation of the asymptotic expansion of the distribution function. This motivates the following definition.

DEFINITION. *Let e be a $\star$-asymptotic scale. We say that a function f is of class m for e if it admits an asymptotic expansion in the scale e and the following diagram commutes.*

$$\begin{array}{ccccccc} f & \xrightarrow{\mathrm{D}} & f' & \xrightarrow{\mathrm{D}} & \cdots & \xrightarrow{\mathrm{D}} & f^{(m)} \\ \downarrow{p} & & \downarrow{p} & & & & \downarrow{p} \\ p_f & \xrightarrow{\mathcal{D}} & p_{f'} & \xrightarrow{\mathcal{D}} & \cdots & \xrightarrow{\mathcal{D}} & p_{f^{(m)}} \end{array}$$

If m and n are two nonnegative integers with m less then n, a function of class n for a $\star$-asymptotic scale is also of class m for that scale. This leads to the following definition.

DEFINITION. *A function is of class ∞ for a $\star$-asymptotic scale if it is of class m for any positive integer m.*

In order to use this definition in concrete examples, we need a good way to determine that a function is of class m for a given $\star$-asymptotic scale. We do not know a general result which makes easy such determination; but, through the next two propositions and their proofs, we can give some useful indications on how to proceed in particular cases.

PROPOSITION 3.2.2. *Let α and τ be some positive real numbers, and let g be a function analytic in the neighborhood of the origin of the real line. The function $f(t) = t^{-\alpha} g(t^{-\tau})$ is of class ∞ for the $\star$-asymptotic scale $\mathrm{Id}^{-\alpha-i\tau-j}$, $i, j \in \mathbb{N}$. Writing $\sum_{i\in\mathbb{N}} g_i \,\mathrm{Id}^i$ for the Taylor series of g,*

$$f \sim \sum_{i\in\mathbb{N}} g_i \,\mathrm{Id}^{-\alpha-i\tau}.$$

PROOF. Ultimately $f = \sum_{i\in\mathbb{N}} g_i \,\mathrm{Id}^{-\alpha-i\tau}$. This function can be ultimately termwise k times differentiated and

$$f^{(k)} = \sum_{i\in\mathbb{N}} (-\alpha - i\tau)_k \, g_i \,\mathrm{Id}^{-\alpha-i\tau-k}.$$

It follows that f and its derivatives have an asymptotic expansion in the scale $\mathrm{Id}^{-\alpha-i\tau-j}$, $i, j \in \mathbb{N}$. This is a $\star$-asymptotic scale. That f is of class m follows from the representation

$$f^{(k)} = \langle (g_i)_{i\geq 0}, (\mathrm{D}^k \mathrm{Id}^{-\alpha-i\tau})_{i\geq 0}\rangle \qquad \blacksquare$$

PROPOSITION 3.2.3. *Let α and τ be two positive real numbers. Let g be a function analytic in a neighborhood of the origin. The function $f(t) = t^{-\alpha} g(\log^{-\tau} t)$ is of class ∞ for the $\star$-asymptotic scale $\mathrm{Id}^{-\alpha} \log^{-i\tau-j}$, $i, j \in \mathbb{N}$. For this scale, $\mathcal{D} = 0$.*

PROOF. With the same notations as in the proof of Proposition 3.2.2,

$$f = \sum_{i\in\mathbb{N}} g_i \,\mathrm{Id}^{-\alpha} \log^{-i\tau}.$$

Note that $\mathrm{D}(\mathrm{Id}^{-\alpha} \log^{-i\tau-j})$ is regularly varying of index $-\alpha - 1$. This implies that $\mathcal{D}$ is 0. Moreover, for any i, j and k positive integers, $f^{(k)} = o(\mathrm{Id}^{-\alpha} \log^{-i\tau-j})$, so that $p_{\mathrm{D}^k f} = 0$. It follows that $\mathcal{D}^k p_f = p_{\mathrm{D}^k f} = 0$. It is easy to check that the asymptotic scale is closed under the multiplication operator. $\blacksquare$

3.3. Tail calculus: From Laplace characters to linear algebra.

The purpose of this subsection is to show how we can now derive tail expansions for weighted convolutions in a very effective way.

Assume that we start with a distribution function F such that both $\overline{F}$ and $\overline{M_{-1}F}$ are of class m for an asymptotic scale $e = (e_i)_{i \in I}$ say. That is, among other things, that for some coefficients $p_{\overline{F},i}$ and $p_{\overline{M_{-1}F},i}$,

$$\overline{F} \sim \sum_{i \in I} p_{\overline{F},i} e_i \qquad \text{and} \qquad \overline{M_{-1}F} \sim \sum_{i \in I} p_{\overline{M_{-1}F},i} e_i \,.$$

Write N for the cardinality of I. Attached to the $\star$-asymptotic scale is a $N \times N$ matrix $\mathcal{D}$. The correspondence $\mathrm{D} \mapsto \mathcal{D}$ defines an isomorphism from $\mathbb{R}_m[\mathrm{D}]$ to the space of $N \times N$ matrices. This isomorphism allows one to identify the Laplace character $L_{K,m}$ with the matrix

$$\mathcal{L}_K = \sum_{0 \leqslant j \leqslant m} \frac{(-1)^j}{j!} \mu_{K,j} \mathcal{D}^j \,.$$

Note that this matrix depends on the order m of the Laplace character as well as on the $\star$-asymptotic scale chosen. The notation does not show this dependence. The map $F \mapsto \mathcal{L}_F$ is a linear representation of the convolution algebra, whose dimension is the cardinality of I. In particular $\mathcal{L}_F \mathcal{L}_G = \mathcal{L}_{F \star G}$. Theorem 2.5.1 implies that the following diagram commutes.

$$\begin{array}{ccc} (F, G) & \xrightarrow{\quad \star \quad} & F \star G \\ \downarrow & & \downarrow {\scriptstyle (p, \mathcal{L})} \\ (p_{\overline{F}}, p_{\overline{G}}, \mathcal{L}_F, \mathcal{L}_G) & \longrightarrow & (\mathcal{L}_F p_{\overline{G}} + \mathcal{L}_G p_{\overline{F}}, \mathcal{L}_F \mathcal{L}_G) \end{array}$$

Practically, the diagrams say that for a distribution function of class m for a $\star$-asymptotic scale, asymptotic expansions for weighted convolution can be done by manipulating finite dimensional vectors and matrices. This algebraic calculus, because of the commutativity of the diagram, is a calculus on tails of distributions, hence the name 'tail calculus'. It is the key to effective computation.

In particular, using the existence of inverse for the Laplace characters, the asymptotic expansion of Theorem 2.5.1 is given by

$$\begin{aligned} p_{\overline{F_C}}^{(k)} &= \sum_{c \in C} \mathcal{L}_{F_C} \mathcal{L}_{M_c F}^{-1} \mathcal{D}^k \mathcal{M}_c \big(\mathbb{1}\{c > 0\} p_{\overline{F}} + \mathbb{1}\{c < 0\} p_{\overline{M_{-1}F}} \big) \\ &= \mathcal{L}_{F_C} \bigg(\sum_{c : c > 0} \mathcal{L}_{M_c F}^{-1} \mathcal{D}^k \mathcal{M}_c p_{\overline{F}} + \sum_{c : c < 0} \mathcal{L}_{M_c F}^{-1} \mathcal{D}^k \mathcal{M}_c p_{\overline{M_{-1}F}} \bigg) \,. \end{aligned}$$
(3.3.1)

This expresses the asymptotic expansion as sums and products of matrices. As we did in Barbe and McCormick (2004), this method of calculation is suitable for computer implementation. It allows one to automatize asymptotic expansion for this type of weighted heavy tail convolutions. This will be illustrated and explained further in the next subsection.

To close this subsection, we comment on the main difference between the approximation formula in Theorem 2.5.1 and that provided by (3.3.1). In order to use (3.3.1), one needs to define an asymptotic scale. The formula in Theorem 2.5.1 does not rely on any scale. If one is interested in approximating $\overline{F}_C$ up to $o(\mathrm{Id}^{-m}\overline{F})$, Theorem 2.5.1 provides the result, independently of an asymptotic scale. Perhaps to insist on that point, we will see in the next section that, for the log-gamma distribution, we cannot find a good scale which allows one to expand simultaneously the tail $\overline{F}$ and that of $\overline{F}_C$; in that situation, Theorem 2.5.1 yields an expansion far more precise than what (3.3.1) can provide. Finally, in subsection 3.5 and in section 4, we will discuss some problems where the formula in Theorem 2.5.1 is far more useful than that given by (3.3.1).

3.4. Examples.

In the next two examples, we work out the details and show how to obtain expansions with as many terms as the integrability condition $m < \alpha$ and computer memory allow. The third and last example in this subsection contains the proof of the last statement of subsection 3.1

Distributions with asymptotic expansion of the form $\sum_{i \geqslant 0} a_i t^{-\alpha_i}$. The Hall-Weissman (1997) distributions are defined by

$$\overline{F}(t) = at^{-\alpha} + bt^{-\beta},$$

with $\alpha < \beta$ and for t large enough. They have an expansion of the form $\sum_{i \geqslant 0} a_i t^{-\alpha_i}$ and are of class ∞ for the $\star$-asymptotic scale $\mathrm{Id}^{-\alpha-i}$, $\mathrm{Id}^{-\beta-i}$ for $i \in \mathbb{N}$.

The Burr distributions with positive parameters τ and γ are defined by

$$\overline{F}(t) = (1 + t^\tau/\beta)^{-\gamma}, \qquad t \geqslant 0.$$

When γ is 1, it is also called the log-logistic distribution. Proposition 3.2.2 shows that it has an asymptotic expansion and is of class ∞ for the $\star$-asymptotic scale $\mathrm{Id}^{-\tau\gamma-i\tau-j}$, $i,j \in \mathbb{N}$. In particular, writing $\overline{F}(t) = \beta^\gamma t^{-\gamma\tau}(1 + \beta t^{-\tau})^{-\gamma}$, we obtain

$$\overline{F}(t) \sim \beta^\gamma \sum_{k \geqslant 0} (-1)^k \beta^k \frac{\Gamma(\gamma + k)}{k! \Gamma(\gamma)} t^{-\tau\gamma - k\tau}. \qquad (3.4.1)$$

This expansion is of the form $\sum_{i \geqslant 0} a_i t^{-\alpha_i}$.

The Fréchet distribution is defined by $\overline{F}(t) = 1 - \exp(-t^{-\alpha})$. Proposition 3.2.2 shows that it has an asymptotic expansion and is of class ∞ for the $\star$-asymptotic scale $\mathrm{Id}^{-\alpha i - j}$, $i, j \in \mathbb{N}$. Explicitly,

$$\overline{F}(t) \sim \sum_{k \geqslant 1} \frac{(-1)^{k+1}}{k!} t^{-\alpha k}.$$

Again, this expansion has the form $\sum_{i\geqslant 0} a_i t^{-\alpha_i}$.

So, it is of some interest here to consider an increasing sequence of real numbers $0 < \alpha_0 < \ldots < \alpha_p$, and investigate expansions for weighted convolutions of distribution functions having an asymptotic expansion in the scale $t^{-\alpha_i}$, $0 \leqslant i \leqslant p$. For a more comprehensive list of heavy tailed distributions with examples of their use in modeling, particularly for insurance risk data, we refer to Beirlant, Teugels and Vynckier (1996), Embrechts, Klüppelberg and Mikosch (1997) and Rolski, Schmidli, Schmidt and Teugels (1999).

Before proceeding, we need to note several things. There is of course no loss of generality in assuming that the coefficient of $t^{-\alpha_0}$ in the tail expansion of $\overline{F}$ does not vanish. Hence, we can assume that $\overline{F}(t) \asymp t^{-\alpha_0}$. We will apply Theorem 2.5.1 with $k = 0$. This requires $m < \alpha_0$; and so, up to truncating the sequence, we can also assume that $\alpha_p \leqslant \alpha_0 + m < 2\alpha_0$. The next thing to notice is that the asymptotic scale $(t^{-\alpha_i})_{0\leqslant i\leqslant p}$ may not be a $\star$-asymptotic scale. Indeed, it is closed under differentiation if and only if for any integer $0 \leqslant i \leqslant p$ and any nonnegative integer k for which $\alpha_i + k \leqslant \alpha_p$, there exists a j such that $\alpha_i + k = \alpha_j$. If this is not the case, we enlarge our collection of α_j's by adding the missing ones recursively. Thus, we assume without loss of generality that $(t^{-\alpha_i})_{0\leqslant i\leqslant p}$ is a $\star$-asymptotic scale.

Now, let us consider the functions $e_i(t) = t^{-\alpha_i}$, $0 \leqslant i \leqslant p$. They form the basis of a finite dimensional vector space, isomorphic to $\mathbb{R}^{p+1}$. Write ϵ_i for the unit vector in $\mathbb{R}^{p+1}$ whose coordinates are all 0 except that of index i (recall that our indexing starts at 0, not the usual 1). Since $De_i = -\alpha_i e_j$ with j defined by $\alpha_j = \alpha_i + 1$, the derivative is identified with the matrix $\mathcal{D}$ determined by

$$\langle \epsilon_i, \mathcal{D}^{\mathrm{t}} e\rangle = \begin{cases} -\alpha_i e_j & \text{if } \alpha_j = \alpha_i + 1 \leqslant \alpha_p, \\ 0 & \text{otherwise.} \end{cases}$$

Since $\mathcal{D}$ maps ϵ_i into the space spanned by $\epsilon_{i+1},\ldots,\epsilon_p$, it is a nilpotent matrix.

Next, since $\mathcal{M}_c e_i = c^{\alpha_i} e_i$, the matrix $\mathcal{M}_c$ is the diagonal matrix $\mathrm{diag}(c^{\alpha_i})_{0\leqslant i\leqslant p}$.

To show how this works practically, we consider the Burr distribution with $\gamma = 10$ and $\tau = 3/2$ say (other values would work just as well). Then $\alpha_0 = \gamma\tau = 15$, so that Theorem 2.5.1 allows for a 14 terms expansion of the weighted convolution. However, we limit the expansion by chosing $m = 4$ in Theorem 2.5.1, though it will be clear at the end that to do so is motivated not by any difficulty in obtaining the terms in such high-order expansion, but rather by concern with the space and display of such expansion. The `Maple` code that we used to implement the formal calculations is given in the appendix.

So we choose m to be 4. Display (3.4.1) shows that $\overline{F}$ has an asymptotic expansion in the scale $t^{-15-3i/2}$, $i \geqslant 0$. Since we want 4 terms and $t^{-4}\overline{F}(t) \asymp t^{-19}$, we can restrict the range of i's to $15 + 3i/2 \leqslant 19$, that is $i \leqslant 2$. The expansion of $\overline{F}$ is then

$$\overline{F}(t) = \beta^{10} t^{-15} - 10\beta^{11} t^{-33/2} + 55\beta^{12} t^{-18} + o(t^{-19}). \qquad (3.4.2)$$

Thus, we need to consider the functions t^{-15}, $t^{-15-3/2}$ and t^{-18}. This family is not closed under differentiation modulo terms in $o(t^{-19})$. To close it, we need to add the derivative of t^{-15}, that is up to a multiplicative constant, t^{-16}; but then we need to add the derivative of t^{-16} as well, that is to add t^{-17}. Similarly, we must add t^{-19}. Next we also need to add the derivative of $t^{-15-3/2}$, which is, up to scale, $t^{-17-1/2}$, and its derivative, up to scale, $t^{-18-1/2}$. This process leads to the $\star$-asymptotic scale $(e_i)_{0 \leqslant i \leqslant 7}$ corresponding to the following α_i's:

$$\alpha_0 = 15 < 16 < 16 + 1/2 < 17 < 17 + 1/2 < 18 < 18 + 1/2 < 19.$$

Since we have eight α_i's, we need to work in the space $\mathbb{R}^8$.

Display (3.4.2) shows that

$$\overline{F} = \beta^{10} e_0 - 10\beta^{11} e_2 + 55\beta^{12} e_5 + o(e_7).$$

So, the vector $p_{\overline{F}}$ in $\mathbb{R}^8$ is simply

$$p_{\overline{F}} = (\beta^{10}, 0, -10\beta^{11}, 0, 0, 55\beta^{12}, 0, 0).$$

Since $De_0 = -15e_1$ and $De_i = -\alpha_i e_{i+2}$ for i at least 1, the matrix that corresponds to the differentiation operator is defined by

$$\mathcal{D} = \begin{pmatrix} 0 & & & & & & & \\ -15 & 0 & & & & & & \\ 0 & 0 & 0 & & & & & \\ & -16 & 0 & 0 & & & 0 & \\ & & -16.5 & 0 & 0 & & & \\ & & & -17 & 0 & 0 & & \\ & 0 & & & -17.5 & 0 & 0 & \\ & & & & & -18 & 0 & 0 \end{pmatrix}.$$

Moreover, since $\mathcal{M}_c e_k = c^{\alpha_k} e_k$, the multiplication operator is represented by

$$\mathcal{M}_c = \operatorname{diag}(c^{\alpha_k})_{0 \leqslant k \leqslant 7}.$$

A Laplace character $L_{K,4}$ is identified with the matrix

$$\mathcal{L}_K = \sum_{0 \leqslant j \leqslant 4} \frac{(-1)^j}{j!} \mu_{K,j} \mathcal{D}^j.$$

This shows that $\mathcal{L}_{\mathcal{M}_c F}$ is the matrix

$$\begin{pmatrix} 1 & & & & & & & \\ 15c\mu_1 & 1 & & & & & & \\ 0 & 0 & 1 & & & & & \\ 120c^2\mu_2 & 16c\mu_1 & 0 & 1 & & & 0 & \\ 0 & 0 & \frac{33}{2}c\mu_1 & 0 & 1 & & & \\ 680c^3\mu_3 & 136c^2\mu_2 & 0 & 17c\mu_1 & 0 & 1 & & \\ 0 & 0 & \frac{1155}{8}c^2\mu_2 & 0 & \frac{35}{2}c\mu_1 & 0 & 1 & \\ 3060c^4\mu_4 & 816c^3 & 0 & 153c^2\mu_2 & 0 & 18c\mu_1 & 0 & 1 \end{pmatrix}.$$

3.4. EXAMPLES

Finally, $\overline{F}_C$ has an expansion whose coefficients are given by

$$\mathcal{L}_{F_C} \sum_{c \in C} \mathcal{L}_{M_c F}^{-1} \mathcal{M}_c p_{\overline{F}}.$$

We formally compute the vector $\mathcal{L}_{M_c F}^{-1} \mathcal{M}_c p_{\overline{F}}$. Assuming that the c's are positive, summing the vectors $\mathcal{L}_{M_c F}^{-1} \mathcal{M}_c p_{\overline{F}}$ leads to expressions involving $C_s = \sum_{c \in C} c^s$. The matrix $\mathcal{L}_G$ involves moments of $\overline{F}_C$ which in turn are expressed in terms of moments of F and other expressions involving the terms C_s.

We can then compute the matrix for the Laplace character associated to F_C. We then do a formal matrix multiplication. To write the final result recall the notation $C_{p;q}$ for $C_{p+q} - C_p C_q$. We also write κ_3 for the third central moment of F, that is for $E(X - \mu_{F,1})^3$; and κ_4 for the fourth one, $E(X - \mu_{F,1})^4$. Our results then show that $P\{\langle c, X \rangle > t\}$ has asymptotic expansion $\sum_{0 \leqslant i \leqslant 7} q_i e_i$ with

$$q_0 = \beta^{10} C_{15},$$
$$q_1 = -15\beta^{10} \mu_1 C_{15;1},$$
$$q_2 = -10\beta^{11} C_{33/2},$$
$$q_3 = 120\beta^{10} \mu_1^2 (C_{16;1} - C_1 C_{15;1}) - 120\beta^{10} \sigma^2 C_{15;2},$$
$$q_4 = 165\beta^{11} \mu_1 C_{33/2;1},$$
$$q_5 = -680\beta^{10} \mu_1^3 (C_{17;1} - 2C_1 C_{16;1} + C_1^2 C_{15;1})$$
$$\qquad + 2040\beta^{10} \sigma^2 \mu_1 (C_{17;1} - C_2 C_{15;1} - 680\beta^{10} \kappa_3 C_{15;3} + 55\beta^{12} C_{18}),$$
$$q_6 = \frac{5775}{4}\beta^{11}\left(-\mu_1^2(C_{35/2;1} - C_1 C_{33/2;1}) + \sigma^2 C_{33/2;2}\right)$$
$$q_7 = 3060\beta^{10} \mu_1^4 (C_{18;1} - 3C_1 C_{17;1} - 3C_1^2 C_{16;1} - C_1^3 C_{15;1})$$
$$\qquad - 18360\beta^{10} \sigma^2 \mu_1^2 (C_{18;1} - C_1 C_{17;1} - C_2 C_{16;1} - C_1 C_2 C_{15;1})$$
$$\qquad - 9180\beta^{10} \sigma^4 (2C_{17;2} - C_{15} C_{2;2})$$
$$\qquad + 12240\beta^{10} \kappa_3 \mu_1 (C_{18;1} - C_3 C_{15;1})$$
$$\qquad - 990\beta^{12} \mu_1 C_{18;1} - 3060\beta^{10} \kappa_4 C_{15;4}.$$

It should be clear now that Theorem 2.5.1 is not a pure abstraction and that the algebraic nature of the Laplace character is what makes such a computation possible.

The log-gamma distribution. Recall that X has a log-gamma distribution with parameter (λ, α) if its density is given by

$$f(x) = \frac{\alpha^\lambda}{\Gamma(\lambda)} (\log x)^{\lambda-1} x^{-\alpha-1}, \quad x \geqslant 1.$$

This is equivalent to saying that $X = \exp(Z/\alpha)$ where Z has a standard Gamma distribution with parameter λ. Applying Proposition 3.2.3 the log-gamma distribution has an asymptotic expansion and is of class ∞ for the $\star$-asymptotic scale $e_k = \mathrm{Id}^{-\alpha} \log^{-\lambda-1-k}$, $k \in \mathbb{N}$. Successive integrations by parts show that

$$P\{Z > t\} \sim \frac{e^{-t}}{\Gamma(\lambda)} \sum_{k \geqslant 0} (\lambda - 1)_k t^{\lambda-1-k}.$$

The distribution function F of X is given by $P\{Z \leqslant \alpha \log x\}$. Thus, it has expansion
$$\overline{F} \sim \frac{1}{\Gamma(\lambda)} \sum_{k \geqslant 0} (\lambda - 1)_k \alpha^{\lambda - 1 - k} e_k \,.$$

As was seen in Proposition 3.2.3, the matrix $\mathcal{D}$ is 0 for this scale. To determine the matrices $\mathcal{M}_c$, we write
$$e_j(t/c) = c^\alpha t^{-\alpha} (\log t)^{\lambda - 1 - j} \left(1 - \frac{\log c}{\log t}\right)^{\lambda - 1 - j}$$
$$\sim \sum_{k \geqslant 0} \frac{(\lambda - 1 - j)_k}{k!} (-1)^k c^\alpha (\log c)^k e_{j+k}(t) \,.$$

Hence, the matrix $\mathcal{M}_c$ is the lower triangular matrix defined by
$$(\mathcal{M}_c)_{i,j} = \begin{cases} \frac{(\lambda - 1 - j)_{i-j}}{(i-j)!} (-1)^{i-j} c^\alpha (\log c)^{i-j} & \text{if } i \geqslant j \geqslant 0, \\ 0 & \text{otherwise.} \end{cases}$$

Note that we indexed our matrix so that the first row and column are labeled 0.

Since $\mathcal{D}$ vanishes, the Laplace character of F is the identity. So the formula in Theorem 2.5.1 yields for any α more than 1,
$$p_{\overline{F_C}} = \sum_{c \in C} \mathcal{M}_c p_{\overline{F}}$$

with
$$p_{\overline{F}} = \frac{\alpha^{\lambda - 1}}{\Gamma(\lambda)} \left(1, \frac{\lambda - 1}{\alpha}, \frac{(\lambda - 1)_2}{\alpha^2}, \ldots, \frac{(\lambda - 1)_k}{\alpha^k}\right).$$

Writing $(C \log C)_{r,s}$ for $\sum_{c \in C} c^r (\log c)^s$, we obtain for instance the asymptotic expansion
$$\overline{F}_C \sim \sum_{j \geqslant 0} q_j e_j$$

with
$$q_j = \frac{\alpha^{\lambda - 1}}{\Gamma(\lambda)} \sum_{0 \leqslant i \leqslant j} \frac{(\lambda - 1)_{j-i}}{\alpha^{j-i}} \frac{(\lambda - 1 - j + i)_i}{i!} (-1)^i (C \log C)_{\alpha, i} \,.$$

The interesting feature of this example is that for any m, the identification $\mathcal{L}_F$ of the m-th Laplace character of F is the identity. A consequence is that in this asymptotic scale, increasing m in Theorem 2.5.1 does not yield extra terms compared to taking m to be 0. This is entirely due to the choice of the asymptotic scale. However, the expansion given in Theorem 2.5.1 provides in fact a better estimate. The reason is that the asymptotic scale in Theorem 2.5.1 is adapted to the distribution function F and its derivatives. It is not a scale given a priori on which $\overline{F}$ is expanded.

This situation is very similar to the fact that the normal tail
$$\overline{\Phi}(t) = \int_t^\infty \frac{e^{-x^2/2}}{\sqrt{2\pi}} \, dx$$

has a one term expansion in the scale $\overline{\Phi}^k$, $k \geqslant 1$, say, but it has an expansion given by a divergent series in the asymptotic scale $t^{-k}e^{-t^2/2}$, $k \geqslant 0$.

A degenerate case. For the log-gamma distribution, we saw that we can find an asymptotic scale such that Theorem 2.5.1 with m equal 0 provides as many terms as we like. In the example to be developed now, we show that the reverse situation may occur; that is, due to rather exceptional cancellations, there are distributions such that Theorem 2.5.1 provides only 1 term. This was the statement ending subsection 3.1. For simplicity we will show this only when the c_i's are nonnegative. It is conceptually easy to adapt the proof if this does not hold.

So, let us first choose m, as large as we want. Let α be larger than m, and let $C = (c_i)_{i \in \mathbb{Z}}$ be a sequence of nonnegative numbers such that $N_{\alpha,\gamma,\omega}(C)$ is finite. We consider the $\star$-asymptotic scale $e_i(t) = t^{-\alpha-i}$, $0 \leqslant i \leqslant m$. As we have seen in the first example of this subsection, the derivative D is identified with the matrix $\mathcal{D}$ whose entries are

$$\mathcal{D}_{i,j} = \begin{cases} -\alpha - j & \text{if } i = j+1, \\ 0 & \text{otherwise.} \end{cases}$$

Moreover, for any c positive,

$$\mathcal{M}_c = \text{diag}(c^\alpha, c^{\alpha+1}, \ldots, c^{\alpha+m}).$$

As before, the m-th Laplace character of F is identified with the matrix

$$\mathcal{L}_{F,m} = \sum_{0 \leqslant j \leqslant m} \frac{(-1)^j}{j!} \mu_{F,j} \mathcal{D}^j.$$

Let $p = (p_0, \ldots, p_m)$ be the coefficients of the expansion of $\overline{F}$ in the asymptotic scale e_i, $0 \leqslant i \leqslant m$. Then, the coefficients of $\overline{F}_C$ in this asymptotic scale are given by

$$q = \mathcal{L}_{F_C,m} \sum_{c \in C} \mathcal{L}_{\mathcal{M}_c F,m}^{-1} \mathcal{M}_c p.$$

So, it suffices to prove that we can find F such that $q = e_0$. Note that there is no loss of generality in assuming all the first m moments of F fixed, because fixing these moments does not put any restriction on the vector p. Hence the first m moments of F_C are fixed. In other words both the Laplace characters of F and F_C can be taken fixed. We already know that $\mathcal{L}_{F_C,m}$ is invertible for it is a lower triangular matrix with all diagonal terms equal to 1. We obtain an expression for the inverse of $\mathcal{L}_{\mathcal{M}_c F}$. Specifically, we define the nilpotent matrix $\mathcal{N} = \text{Id} - \mathcal{L}_{\mathcal{M}_c F,m}$ and have

$$\mathcal{L}_{\mathcal{M}_c F,m}^{-1} = (\text{Id} - \mathcal{N})^{-1} = \sum_{0 \leqslant j \leqslant m} \mathcal{N}^j$$

is also lower triangular, with its diagonal elements all equal to 1. Consequently, the diagonal elements of $\mathcal{L}_{\mathcal{M}_c F,m}^{-1} \mathcal{M}_c$ are those of $\mathcal{M}_c$. Hence the matrix $\sum_{c \in C} \mathcal{L}_{\mathcal{M}_c F,m}^{-1} \mathcal{M}_c$ is lower triangular, with diagonal elements $(C_\alpha, C_{\alpha+1}, \ldots, C_{\alpha+m})$. It is invertible. Therefore it is possible to find p with positive e_0 component such that q is e_0. This shows the existence of a distribution function F such that the formula in Theorem 2.5.1 yields only a 1 term expansion.

This discussion is particularly relevant to the problem of ascertaining second-order regular variation for infinite-order moving averages. If an m-term expansion shows that, for example, $\overline{F}_C \sim \mathrm{Id}^{-\alpha} + \mathrm{Id}^{-\alpha-m}$, then $\overline{F}_C$ is second-order regularly varying; but this would not be revealed until an m-terms expansion was calculated. Thus, we see that second-order regular variation for linear processes is not a second-order expansion question; rather, it is a higher-order expansion question. We discuss this problem in detail in the next subsection.

3.5. Two-term expansion and second-order regular variation.

Motivated by probabilistic and statistical applications, consider the following problem: If $\overline{F}$ is regularly varying of index $-\alpha$ with remainder, what is an asymptotic equivalent of $\overline{F}_C - C_\alpha \overline{F}$?

In this section, we will first explain this problem and the terminology used, and then show that Theorem 2.5.1 sheds an interesting light on the matter.

Recall that $\overline{F}$ is regularly varying with index $-\alpha$ and remainder, if there exist functions $k(\cdot)$ and $g(\cdot)$, with g tending to 0 at infinity, such that

$$\frac{\overline{F}(\lambda t)}{\overline{F}(t)} - \lambda^{-\alpha} \sim \lambda^{-\alpha} k(\lambda) g(t) \qquad (3.5.1)$$

as t tends to infinity — see, Bingham, Goldie and Teugels (1989, §3.12). Note that this relation does not change if we multiply g and divide k by the same constant. If this relation holds, then g must be regularly varying with nonpositive index ρ and, up to a possible multiplication of g by a constant, necessarily

$$k(\lambda) = \begin{cases} (\lambda^\rho - 1)/\rho & \text{if } \rho < 0, \\ \log \lambda & \text{if } \rho = 0. \end{cases}$$

We then write $\overline{F} \in 2RV(-\alpha, g)$. Unless otherwise specified, we assume for simplicity that the elements of C are nonnegative. The problem mentioned can be rephrased as: assuming (3.5.1), we know that $\overline{F}$ is regularly varying with index $-\alpha$, hence that

$$\lim_{t \to \infty} \overline{F}_C(t)/\overline{F}(t) = C_\alpha \,.$$

What is the exact rate of convergence in this limit? One would indeed think that adding one term to the regular variation as in (3.5.1) brings one more term in the asymptotic expansion for $\overline{F}_C$. The examples in subsection 3.1 show that this belief is not correct in general.

Since the problem is motivated by applications in time series where it is natural to suppose that F is centered, we assume that $\mu_{F,1}$ vanishes — the following discussion can easily be modified if this first moment does not vanish. In order to apply Theorem 2.5.1 with $k = 0$ and $m = 2$, we assume that $\overline{F}$ is smoothly varying of order larger than 2 and that α is larger than 2. We calculate the second Laplace character $L_{F_{C\backslash c},2} = \mathrm{Id} + (1/2)\mu_{F,2}(C_2 - c^2)\mathrm{D}^2$. Then Theorem 2.5.1, with $m = 2$

and $k=0$, and the fact that $\overline{F}'' \sim \alpha(\alpha+1)\mathrm{Id}^{-2}\overline{F}$ at infinity, yield

$$\overline{F}_C = \sum_{c\in C} \overline{M_c F} - \frac{\alpha(\alpha+1)}{2}C_{\alpha;2}\mu_{F,2}\mathrm{Id}^{-2}\overline{F} + o(\mathrm{Id}^{-2}\overline{F})$$

at infinity. Consequently,

$$\frac{\overline{F}_C}{\overline{F}}(t) - C_\alpha = \sum_{c\in C}\Big(\frac{\overline{F}(t/c)}{\overline{F}(t)} - c^\alpha\Big) - \frac{\alpha(\alpha+1)}{2}C_{\alpha;2}\mu_{F,2}t^{-2} + o(t^{-2}).$$

The global Potter type bounds of Theorem 3.1.3 in Bingham, Goldie and Teugels (1989) and (3.5.1) imply

$$\frac{\overline{F}_C}{\overline{F}}(t) - C_\alpha = g(t)\sum_{c\in C} c^\alpha k(1/c)\big(1+o(1)\big) - \frac{\alpha(\alpha+1)}{2}C_{\alpha;2}\mu_{F,2}t^{-2} + o(t^{-2}).$$

Consequently, one sees that the second-order term depends on the behavior of $t^2 g(t)$ at infinity. If $\lim_{t\to\infty} t^2 g(t) = \infty$, then

$$\frac{\overline{F}_C}{\overline{F}}(t) - C_\alpha \sim g(t)\sum_{c\in C} c^\alpha k(1/c)$$

while if $\lim_{t\to\infty} t^2 g(t) = 0$ then

$$\frac{\overline{F}_C}{\overline{F}}(t) - C_\alpha \sim -\frac{\alpha(\alpha+1)}{2}C_{\alpha;2}\mu_{F,2}t^{-2},$$

with the usual convention that if the constant $C_{\alpha;2}$ in the right hand side vanishes, then the right hand side should be read as $o(t^{-2})$. If $g(t) \sim at^{-2}$, then ρ is -2 and we obtain

$$\frac{\overline{F}_C}{\overline{F}}(t) - C_\alpha \sim \frac{1}{2t^2}\Big(-a(C_{\alpha+2} - C_\alpha) - \alpha(\alpha+1)C_{\alpha;2}\mu_{F,2}\Big).$$

If the constant in the second-order terms vanishes (which generically does not happen), then if we can, we need to add one more term when applying Theorem 2.5.1. But one should be careful that the rate of convergence of $t^2 g(t)$ to its limit may cancel the extra term added. If cancellation occurs, then more terms are needed, and so on. It is therefore not clear that second-order regular variation provides the right framework for studying second-order expansions of $\overline{F}_C$. In particular, for some exceptional sequences of constants and some distributions, higher-order regular variation will be needed to obtain the exact second order. We also would like to point out that smooth variation of finite order is far easier to check than second-order regular variation, and that it holds for most — if not all — heavy tail distributions used in practical applications.

We conclude this section with a somewhat more general result to illustrate how weights of arbitrary signs appear in the expansion. If X is a random variable with distribution function F, we write F_* for the distribution function of $|X|$. Thus, on the nonnegative half-line, $\overline{F}_* = \overline{M_{-1}F} + \overline{F}$. In the statistical literature dealing with regular variation for the upper and lower tails of distributions, it is customary to replace (3.5.1) by a second-order regular variation assumption on $\overline{F}_*$, that is, with the same notation as in (3.5.1),

$$\frac{\overline{F}_*(\lambda t)}{\overline{F}_*(t)} - \lambda^{-\alpha} \sim \lambda^{-\alpha} k(\lambda) g(t),$$

and a tail balancing condition with remainder, that is for some nonnegative p at most 1,

$$\overline{F} = p\overline{F}_* + o(g\overline{F}_*)$$

at infinity. Set $q = 1 - p$ and

$$\kappa_C(\lambda) = \sum_{c \in C} (\lambda/|c|)^{-\alpha} k(\lambda/|c|) \big(p\mathbb{1}\{c > 0\} + q\mathbb{1}\{c < 0\} \big).$$

Define the constants

$$C_{+,\alpha} = \sum_{\substack{c \in C \\ c > 0}} c^\alpha \qquad \text{and} \qquad C_{-,\alpha} = \sum_{\substack{c \in C \\ c < 0}} |c|^\alpha.$$

Furthermore, set

$$C_{*,\alpha} = pC_{+,\alpha} + qC_{-,\alpha}$$

and

$$C_{*,\alpha;1} = p(C_{+,\alpha+1} - C_1 C_{+,\alpha}) - q(C_{-,\alpha+1} + C_1 C_{-,\alpha}).$$

Assume that α is larger than 1 and that F satisfies the assumptions of Theorem 2.5.1 with $m = 1$ and $k = 0$. Finally, assume also that $a = \lim_{t \to \infty} tg(t)$ exists, possibly infinite. Then,

$$\frac{\overline{F}_C(\lambda t)}{\overline{F}_*(t)} - C_{*,\alpha} \lambda^{-\alpha} \sim \begin{cases} \big(a\kappa_C(\lambda) - \lambda^{-\alpha-1} \alpha \mu_{F,1} C_{*,\alpha;1}\big) t^{-1} & \text{if } a \text{ is finite,} \\ \kappa_C(\lambda) g(t) & \text{if } a \text{ is infinite.} \end{cases}$$

We remark that the above results extend Theorem 3.2.III in Geluk, De Haan, Resnick and Stărică (1997).

3.6. Some open questions.

Formula (3.1.3) provides generically a two terms expansion when the mean $\mu_{F,1}$ vanishes. For this expansion to be valid, we need a variance in order to define the second Laplace character. Consequently, we do not know what a two terms expansion is when the distribution pertaining to F is centered and has infinite variance; that is essentially in the range α between 1 and 2.

When α is less than 1, the mean does not exists, and we only have the classical equivalence $\overline{F}_C \sim \sum_{c \in C} \overline{M_c F}$. Some examples in subsection 3.1 and the log-gamma distribution studied in subsection 3.4 suggest that in some instances this equivalent may still provide two terms or more depending on the asymptotic scale used.

When α is less than 1 and F is concentrated on the nonnegative half-line with smoothly varying tail of index α and order more than 1, Theorems 2.5 and 2.6 of Barbe and McCormick (2005) imply a result for some finite convolutions. Specifically, define

$$I(\alpha) = 2 \int_0^{1/2} \left((1-y)^{-\alpha} - 1\right)\alpha y^{-\alpha-1}\,\mathrm{d}y + 2^{2\alpha} - 2^{\alpha+1}\,.$$

Then, if the c_i's are nonnegative constants,

$$\overline{\star_{1\leqslant i\leqslant n} M_{c_i} F} = \sum_{1\leqslant i\leqslant n} \overline{M_{c_i} F} + \frac{I(\alpha)}{2}\bigg(\sum_{1\leqslant i,j\leqslant n} c_i^\alpha c_j^\alpha - \sum_{1\leqslant i\leqslant n} c_i^{2\alpha}\bigg)\overline{F}^2\,.$$

This suggests that under suitable conditions, the tail of the distribution of the series $\sum_{i\in\mathbb{Z}} c_i X_i$ should behave like

$$\sum_{i\in\mathbb{Z}} \overline{M_{c_i} F} + \frac{I(\alpha)}{2}(C_\alpha^2 - C_{2\alpha})\overline{F}^2 + o(\overline{F}^2)\,. \tag{3.6.1}$$

The techniques used in Barbe and McCormick (2005) combined with those of the current paper can certainly be used to prove (3.6.1). However, even in the case of a finite number of summands, when α is less than 1, we do not know a good formalism to remove the restriction that the support of the $M_{c_i}F$'s should be in the nonnegative half-line.

We believe that the techniques and formalism developed in the proof of Theorem 2.5.1 of this paper are generally useful for extending a result from finite convolutions to infinite ones. Unfortunately, at the present time, finite convolutions are still difficult to work with.

CHAPTER 4

Applications.

In this section we develop some applications of Theorem 2.5.1 and the tail calculus explained in section 3.3. As the development proceeds, these applications leave room to more and more questions; the last subsection, on implicit renewal equation, only touches on a subject which deserves further consideration.

4.1. ARMA models.

ARMA models are among the most used models in statistical analysis of time series. Yet, very few facts are known on their distributions. The purpose of this subsection is to show that in some circumstances, Theorem 2.5.1 provides some basic information on the marginal distribution. We refer to Brockwell and Davis (1991) for the basic probabilistic and statistical aspects of these models.

To fix notation, we define the backward shift operator B on sequences as follows. A sequence $x = (x_i)_{i \in \mathbb{Z}}$ is mapped under B to the sequence whose i-th element is x_{i-1}. As usual, B^0 is the identity, B^{-1} is the inverse of B, and $B^k = BB^{k-1}$. It then makes sense to consider polynomials in B. Having two polynomials Θ and Φ, and a sequence $\epsilon = (\epsilon_i)_{i \in \mathbb{Z}}$ of independent and identically distributed random variables, an ARMA process $X = (X_i)_{i \in \mathbb{Z}}$ is defined by the relation

$$\Theta(B)X = \Phi(B)\epsilon.$$

For this process to be defined, we assume that Θ has all its roots outside the closed unit disk of the complex plane. We can define $\Theta(B)^{-1}$ by a series expansion. Then, under a mild integrability condition on ϵ, we obtain $X = \Theta(B)^{-1}\Phi(B)\epsilon$. Having a series expansion yields a representation of X as an infinite-order moving average

$$X_i = \sum_{j \in \mathbb{N}} c_j \epsilon_{i-j}.$$

In general it is unknown how to calculate the marginal distribution of the process, that is the distribution of X_0. Since the roots of Θ are outside the closed unit disk, the sequence c_j decreases exponentially fast; see e.g. Brockwell and Davis (1991,§3.1). Therefore, this sequence has finite $N_{\alpha,\gamma,\omega}$-norm, whatever α, γ and ω are in the positive half-line. Theorem 2.5.1 yields immediately an expansion for the tail of the marginal distribution of the process provided the distribution of the innovations ϵ_i has tail smoothly varying of sufficiently large order and index.

To discuss further, let C be the sequence $(c_i)_{i \in \mathbb{Z}}$, so that the marginal distribution of the process is F_C. In order to make the expansion explicit, for any c in C, we need

39

to evaluate the Laplace character $L_{F_{C\setminus c},m}$. This requires computing the moments of $F_{C\setminus c}$, or, equivalently, $E(X_0 - c_i\epsilon_{-i})^k$ for various integers k. We do not know any way to obtain a nice formula for those moments in terms of the polynomials Θ and Φ. In general, one needs to rely on numerical methods, eventually based on the results of Leonov and Shiryaev (1959).

There are however two cases for which explicit calculations may be performed, namely for AR(1) and MA(q) models, that is when either $\Theta(B) = \mathrm{Id} - aB$ and $\Phi(B) = \mathrm{Id}$, or $\Theta(B) = \mathrm{Id}$ and Φ is an arbitrary polynomial of degree q. For instance, in many examples of section 3, the expansions can be expressed with coefficients involving the quantity $C_p = \sum_{i\in\mathbb{Z}} c_i^p$. For an AR(1), with autocorrelation a positive and less than 1,

$$C_p = \sum_{i\geqslant 0} a^{ip} = (1-a^p)^{-1},$$

while for a MA(q) with $\Phi(B) = \sum_{0\leqslant i\leqslant q} \Phi_i B^i$,

$$C_p = \sum_{0\leqslant i\leqslant q} \Phi_i^p.$$

By applying these formulas to the expression obtained in section 3.3, we obtain two terms expansions of the marginal distribution of these processes with an assumption of second-order regular variation for the distribution of the innovations ϵ. This will be used in the next subsection where we develop an application to statistical inference for heavy tail data.

4.2. Tail index estimation.

In heavy tail analysis, a critical parameter to estimate is the index of regular variation when the marginal distribution of the data has a regularly varying tail. In this example we are concerned with observations that follow a causal linear process which we denote $(Y_i)_{i\in\mathbb{N}^*}$. Thus, we assume that there is a sequence of real number c_i's and a sequence of independent and identically distributed random variables X_i's, so that

$$Y_i = \sum_{j\geqslant 0} c_j X_{i-j}, \qquad i \in \mathbb{N}^*.$$

By the first-order result on tail behavior, we have that, if the distribution function of the innovations X_i's has regularly varying tails with index $-\alpha$, then the same is true for the marginal distribution function of the stationary linear process. The statistical problem is to estimate and find a confidence interval for α, based on sample data.

The usual semiparametric procedure is to use the Hill (1973) estimator, defined as follows. Let $|Y|_{i,n}$ be the i-th largest value among $|Y_1|,\ldots,|Y_n|$, so that $|Y|_{n,n} \leqslant \cdots \leqslant |Y|_{1,n}$. Let k_n be an integer between 1 and n. The Hill estimator is

$$\alpha_n = k_n \Big(\sum_{1\leqslant i\leqslant k_n} \log \frac{|Y|_{i,n}}{|Y|_{k_n+1,n}} \Big)^{-1}.$$

For this estimate to be consistent, it is necessary to take points coming from the tail of the distribution, and of course we need enough of them. This is expressed by the conditions

$$\lim_{n\to\infty} k_n/n = 0 \quad \text{and} \quad \lim_{n\to\infty} k_n = \infty. \qquad (4.2.1)$$

To derive a nondegenerate limiting distribution, and therefore obtain asymptotic confidence intervals and tests, further hypotheses need to be imposed. These were obtained by Resnick and Stărică (1997), and until the present paper could not be verified except maybe in some very special situations. To explain what the problem is, we need to list their conditions, and this requires further notation as well as some rather technical consideration.

Let F_* be the distribution function of $|X_i|$. It is assumed that F_* is second-order regularly varying; that is, with the same notation as in section 3.5, F_* belongs to $2RV(-\alpha, g)$ for some regularly varying function g, i.e. (3.5.1) holds with $\overline{F}_*$ in place of $\overline{F}$. Next, it is assumed a tail balancing condition holds, namely that for some p in the closed unit interval and $q = 1 - p$,

$$\overline{F} = p\overline{F}_* + o(g\overline{F}_*) \qquad (4.2.2)$$

and

$$\overline{M_{-1}F} = q\overline{F}_* + o(g\overline{F}_*) \qquad (4.2.3)$$

at infinity. This ensures that both $\overline{F}$ and $\overline{M_{-1}F}$ are second-order regularly varying, with same index $-\alpha$ and same auxiliary rate function g. The next assumption to be made is that F has a density F' which is Lipschitz in mean, that is there exists a positive κ, for which

$$\int_{\mathbb{R}} |F'(x) - F'(x+y)|\, dx \leqslant \kappa y. \qquad (4.2.4)$$

This condition ensures that the linear process is strong mixing.

Note that so far all the conditions are on the unknown distribution of the innovations. In statistical analysis of heavy tailed time series, one takes as one's model assumptions that the innovation distribution satisfies certain properties such as we have listed above. Model assumptions are simply assumed to hold. However, conditions on the marginal distribution of the process may be worrisome. Some may be redundant, i.e. derivable from the assumptions on F, or some may even be inconsistent, i.e. the hypotheses for a theorem may apply to no models. For our particular tail estimation problem, we will elucidate this issue next.

The next condition required is that the marginal distribution function G of the process, that is F_C, satisfies a von Mises condition; this means that it has a density G' and

$$\lim_{t\to\infty} \frac{tG'(t)}{\overline{G}(t)} = \alpha. \qquad (4.2.5)$$

Let G_* be the distribution function of the absolute value $|Y_i|$. It is further assumed that $\overline{G}_*$ is second-order regularly varying of index $-\alpha$, that is, using the notation

introduced after (3.5.1),
$$\overline{G}_* \in 2RV(-\alpha, g_{G_*}) \tag{4.2.6}$$
for some regularly varying g_{G_*}.

Yet, another assumption is that
$$\lim_{n\to\infty} \sqrt{k_n}\, g_{G_*} \circ G_*^{\leftarrow}(1 - k_n/n) = 0. \tag{4.2.7}$$

Assumptions (4.2.5), (4.2.6) and (4.2.7) are rather problematic since they do not involve the distribution function of the innovation, and may potentially put rather stringent conditions on F or k_n. We will investigate that matter after stating Resnick and Stărică's result.

The last assumption is on the sequence k_n, and presents no difficulty, since this sequence is chosen by whomever uses the estimator. This assumption prevents some wild ocillatory behavior of the sequence k_n. It is assumed that
$$\limsup_{n\to\infty} n^{2/3}/k_n < \infty \quad \text{or} \quad \liminf_{n\to\infty} n^{2/3}/k_n > 0. \tag{4.2.8}$$

Set
$$\lambda = \frac{1}{\alpha^2}\left(1 + 2\frac{\sum_{j\geqslant 1}\sum_{k\geqslant 0}|c_k|^\alpha \wedge |c_{j+k}|^\alpha}{\sum_{k\geqslant 0}|c_k|^\alpha}\right).$$

THEOREM 4.2.1. (Resnick, Stărică, 1997). *Let (k_n) be a sequence satifying (4.2.1). Assume that there exists some u more than 1 and some positive constant A such that $|c_i| \leqslant Au^{-i}$ for any nonnegative i and that $E|X_1|^d$ is finite for some positive d less than 1.*
(i) If conditions (4.2.4), (4.2.5) and (4.2.8) hold, then
$$\sqrt{k_n}\left(\alpha_n^{-1} - \frac{n}{k_n}\int_1^\infty P\{\,|Y_1| \geqslant xG_*^{\leftarrow}(1 - k_n/n)\,\}\frac{\mathrm{d}x}{x}\right)$$
has a centered normal limiting distribution, with variance λ.
(ii) If, furthermore, (4.2.6) and (4.2.7) hold, then $\sqrt{k_n}(\alpha_n^{-1} - \alpha^{-1})$ has a normal limiting distribution with mean 0 and variance λ.

Let us now show that Theorem 2.5.1 yields rather simple and explicit conditions which ensure that (4.2.5), (4.2.6) and (4.2.7) hold, and therefore makes Resnick and Stărică's theorem far easier to use. Before stating the result we introduce some conditions on the moving average weights. Recall that in section 3.5, we noted that in general higher-order expansions are required to determine the auxiliary function for an infinite-order weighted average. In order that the derived auxiliary function be determined by only the second-order information for the innovation distribution, certain restrictions must be met. They are encompassed in the following conditions.

Similarly to C_r, we define the notation
$$|C|_r = \sum_{i\in\mathbb{Z}}|c_i|^r.$$

Note that $|C|_r = |c|_r^r$. Recall that $k(\cdot)$ is the function appearing in the definition of second-order regular variation for $\overline{F}_*$ as in (3.5.1). Furthermore, the parameter ρ which appear in the function k is the index of regular variation of the function g. We will use the conditions

$$|C|_\alpha + \rho \sum_{i \in \mathbb{Z}} |c_i|^\alpha k(1/|c_i|) \neq 0, \qquad (4.2.9)$$

and, with a a real number to be fixed later,

$$a|C|_\alpha + a\rho \sum_{i \in \mathbb{Z}} |c_i|^\alpha k(1/|c_i|)$$
$$+ \alpha \rho \mu_{F,1}(p-q)\Big(C_1 \sum_{i \in \mathbb{Z}} |c_i|^\alpha \mathrm{sign}(c_i) - |C|_{\alpha+1}\Big) \neq 0, \quad (4.2.10)$$

as well as

$$a|C|_\alpha + a\rho \sum_{i \in \mathbb{Z}} |c_i|^\alpha k(1/|c_i|) - \alpha(\alpha+1)\mu_{F,2}\big(C_2|C|_\alpha - |C|_{\alpha+2}\big) \neq 0. \quad (4.2.11)$$

We comment on those conditions after the following statement.

PROPOSITION 4.2.2. *Assume that $\overline{F}$ and $\overline{M_{-1}F}$ are smoothly varying of index $-\alpha$ and order more than 1, and that they belong to $2RV(-\alpha, g)$. Assume also that F' ultimately exists and is continuous. Then,*

(i) G obeys the von Mises condition (4.2.5);

(ii) If F'' exists and is Lebesgue integrable, then (4.2.4) holds;

Next, let $\xi = 1$ if $\mu_{F,1}$ does not vanish, and $\xi = 2$ otherwise. Assume furthermore that $\overline{F}$ and $\overline{M_{-1}F}$ are of order more than ξ and that $a = \lim_{t \to \infty} t^\xi g(t)$ exists, possibly infinite.

(iii) The function $\overline{G}_*$ is second-order regularly varying in any of the following three cases:

 case 1. $a = +\infty$ and (4.2.9) holds;
 case 2. a is finite, $\mu_{F,1}$ does not vanish, and (4.2.10) holds;
 case 3. a is finite, $\mu_{F,1}$ vanishes and (4.2.11) holds.

(iv) Moreover,

$$g_{G_*} \asymp \begin{cases} g & \text{if } a \neq 0 \\ \mathrm{Id}^{-\xi} & \text{if } a = 0 \end{cases}$$

and

(v) condition (4.2.7) is equivalent to

$$\begin{cases} \lim_{n \to \infty} \sqrt{k_n}\, g \circ F_*^{\leftarrow}(1 - k_n/n) = 0 & \text{if } a \neq 0, \\ \lim_{n \to \infty} \sqrt{k_n}\, F_*^{\leftarrow}(1 - k_n/n)^{-\xi} = 0 & \text{if } a = 0. \end{cases}$$

REMARK. When $\mu_{F,1}$ does not vanish, the result is obtained by an application of Theorem 2.5.1 with $m = 1$; otherwise it requires $m = 2$. The all but intuitive conditions (4.2.9)–(4.2.11) are not there for technical reasons. If those conditions are not met as specified in the Proposition, then the last statement of the Proposition

— the equivalence with (4.2.7) — does not hold. We would need to use Theorem 2.5.1 with a higher m or higher-order regular variation, which would lead in some cases (but not always) to a different result. Generically, the Proposition covers all cases; however, there are some exceptional ones where it fails. This is the same phenomenon as observed in subsection 3.1, and commented further toward the end of the proof of Proposition 4.2.2. This unfortunate fact commands caution when estimating tail index in time series.

REMARK. To fix the ideas, suppose $\overline{F}(t) \asymp t^{-\alpha}$ and $g(t) \asymp t^{-\beta}$ with β positive and less than ξ. In this situation, a is infinite. Then (4.2.7) is equivalent to $k_n = o(n^{2\beta/(\alpha+2\beta)})$. One sees that the smaller β is, the smaller k_n should be. In crafting a good estimator, one needs to be mindful of such restrictions when using the Hill estimator in a time series context.

PROOF. Let C be the sequence $(c_i)_{i\in\mathbb{Z}}$. Using the notation of Theorem 2.5.1, $G = F_C$.
(i) We first set

$$C_{+,\alpha} = \sum_{i\in\mathbb{Z}\,;\,c_i>0} c_i^\alpha \quad \text{and} \quad C_{-,\alpha} = \sum_{i\in\mathbb{Z}\,;\,c_i<0} (-c_i)^\alpha.$$

Recall the classical first-order equivalence

$$\overline{G} \sim \sum_{i\in\mathbb{Z}} \overline{M_{c_i}F} \sim (pC_{+,\alpha} + qC_{-,\alpha})\overline{F}_*.$$

Applying Theorem 2.5.1 with $k = 1$ and $m = 0$ yields

$$\overline{G}' \sim \sum_{i\in\mathbb{Z}} \overline{M_{c_i}F'} \sim (pC_{+,\alpha} + qC_{-,\alpha})(-\alpha)\mathrm{Id}^{-1}\overline{F}_*.$$

We then deduce (4.2.5).
(ii) follows from the fundamental theorem of calculus as well as Fubini's theorem, upon writing

$$\int |F'(x) - F'(x+y)|\,\mathrm{d}x \leqslant \int\int |F''(x+u)|\,\mathrm{d}x\,\mathbb{1}\{0 \leqslant u \leqslant y\}\,\mathrm{d}u$$
$$\leqslant y|F''|_1.$$

To prove the other statements, we first derive a two terms expansion for $\overline{G}_*$.
We first assume that $\mu_{F,1}$ does not vanish. Since $\overline{G}_*$ coincides with $\overline{G} + \overline{M_{-1}G}$ on the positive half-line, Theorem 2.5.1 and the tail balance conditions (4.2.2), (4.2.3) imply

$$\overline{G}_* = \sum_{c\in C} L_{F_{C\setminus c},1}\overline{M_cF} + \sum_{c\in C} L_{F_{-(C\setminus c)},1}\overline{M_cF} + o(\mathrm{Id}^{-1}\overline{F}_*)$$
$$= \sum_{c\in C}(\overline{M_cF} + \overline{M_{-c}F}) - \mu_{F,1}\sum_{c\in C}(C_1 - c)\mathrm{D}(\overline{M_cF} - \overline{M_{-c}F}) + o(\mathrm{Id}^{-1}\overline{F}_*).$$

We note that on the positive half-line
$$\overline{M_c F} + \overline{M_{-c} F} = \overline{M_{|c|} F_*}.$$

Moreover, because $\overline{F}$ and $\overline{M_{-1}F}$ are smoothly varying of order more than 1, we also have
$$\mathrm{D}\overline{M_c F} \sim -\alpha \mathrm{Id}^{-1} \overline{M_c F}.$$

We then obtain
$$\overline{G_*} = \sum_{c \in C} \overline{M_{|c|} F_*} + \alpha \mu_{F,1} \sum_{c \in C} (C_1 - c)(p-q)\mathrm{sign}(c)\mathrm{Id}^{-1} \overline{M_{|c|} F_*} + o(g)\overline{F_*}.$$

Consequently, we obtain the two terms expansion
$$\frac{\overline{G_*}}{\overline{F_*}} = |C|_\alpha + \sum_{c \in C} |c|^\alpha k(1/|c|) g$$
$$+ \alpha \mu_{F,1}(p-q)\Big(C_1 \sum_{c \in C} \mathrm{sign}(c)|c|^\alpha - |C|_{\alpha+1}\Big)\mathrm{Id}^{-1} + o(\mathrm{Id}^{-1} \vee g).$$

Let us now consider the case where $\mu_{F,1}$ vanishes. Again, Theorem 2.5.1 and the tail balance conditions (4.2.2), (4.2.3) imply
$$\overline{G_*} = \sum_{c \in C} L_{F_{C \setminus c}, 2}(\overline{M_c F} + \overline{M_{-c} F}) + o(\mathrm{Id}^{-2} \overline{F_*})$$
$$= \sum_{c \in C} \overline{M_c F} + \overline{M_{-c} F} + \frac{\mu_{F,2}}{2} \sum_{c \in C} (C_2 - c^2) \mathrm{D}^2 (\overline{M_c F} + \overline{M_{-c} F}) + o(\mathrm{Id}^{-2} \overline{F_*}).$$

Because $\overline{F}$ and $\overline{M_{-1} F}$ are smoothly varying of order more than 2, we also have
$$\mathrm{D}^2 \overline{M_{c_i} F} \sim \alpha(\alpha+1) \mathrm{Id}^{-2} \overline{M_{c_i} F}.$$

Then, up to $o(\mathrm{Id}^{-2} \overline{F_*})$, the tail $\overline{G_*}$ is
$$\sum_{c \in C} \overline{M_{|c|} F_*} + \frac{\mu_{F,2}}{2}\alpha(\alpha+1) \sum_{c \in C} (C_2 - c^2)\mathrm{Id}^{-2} \overline{M_{|c|} F_*}$$
$$= \sum_{c \in C} \overline{M_{|c|} F_*} + \frac{\mu_{F,2}}{2}\alpha(\alpha+1)(C_2 |C|_\alpha - |C|_{\alpha+2})\mathrm{Id}^{-2} \overline{F_*}.$$

Therefore, we have the two terms expansion
$$\frac{\overline{G_*}}{\overline{F_*}} = |C|_\alpha + \sum_{c \in C} |c|^\alpha k(1/|c|)g + \frac{\alpha(\alpha+1)}{2}\mu_{F,2}(C_2 |C|_\alpha - |C|_{\alpha+2})\mathrm{Id}^{-2} + o(\mathrm{Id}^{-2} \vee g).$$

In either the case $\mu_{F,1}$ vanishes or does not vanish, we obtain an expansion of the form
$$\frac{\overline{G_*}}{\overline{F_*}} = U + Vg + W\mathrm{Id}^{-\xi} + o(\mathrm{Id}^{-\xi} \vee g).$$

This implies

$$\frac{\overline{G}_*(\lambda t)}{\overline{G}_*(t)} = \lambda^{-\alpha} + \lambda^{-\alpha} g(t) k(\lambda)(1 + \rho U^{-1} V) + \lambda^{-\alpha} U^{-1} W(\lambda^{-\xi} - 1) t^{-\xi} + o\bigl(t^{-\xi} \vee g(t)\bigr).$$

We now prove the result in case 1. Indeed, if $a = \lim_{t \to \infty} t^\xi g(t)$ is infinite, we have

$$\frac{\overline{G}_*(\lambda t)}{\overline{G}_*(t)} = \lambda^{-\alpha} + \lambda^{-\alpha} g(t) k(\lambda)(1 + \rho U^{-1} V) + o(g).$$

If $1 + \rho U^{-1} V$ does not vanish, which is condition (4.2.9), this implies $\overline{G}_*$ is second-order regularly varying with auxiliary function proportional to g.

The stated equivalence with (4.2.7) follows in this case from the fact that $\overline{G}_*^{\leftarrow}(1 - u) \asymp \overline{F}_*^{\leftarrow}(1 - u)$ as u tends to 0.

This proves statements (iii), (iv) and (v) in case 1.

Note that as we have seen in section 3.1, if $1 + \rho U^{-1} V$ vanishes, we cannot conclude anything without obtaining higher-order expansions, and virtually any auxiliary function may occur, either connected with higher-order regular variation of $\overline{F}_*$ (here we are talking of third-order or even higher order in exceptional cases) or the third terms in the expansion of $\overline{G}_*$ (or higher-order terms in exceptional cases).

Cases 2 and 3 are handled in the same way. ∎

Example. To conclude this example we illustrate the result for a particular distribution. Consider a Student innovation density

$$f(x) = K_\alpha \left(1 + \frac{x^2}{\alpha}\right)^{-(\alpha+1)/2}, \qquad x \in \mathbb{R},$$

where K_α is the normalizing constant and where α is more than 2.

The Student distribution being symmetric, (4.2.2) and (4.2.3) are obvious.

To check the assumptions of Resnick and Stărică's theorem we use Proposition 4.2.2.

It is plain that $\overline{F}$ and $\overline{M_{-1}F}$ are smoothly varying of order at least two, because the second-order derivative of f is regularly varying. Also, f is continuously differentiable with integrable derivative; this establishes that F'' exists, is continuous and integrable.

To prove that $\overline{F}$ is second-order regularly varying, we derive a two terms expansion by writing

$$\begin{aligned}
\overline{F}(t) &= K_\alpha \int_t^\infty x^{-\alpha-1} \alpha^{(\alpha+1)/2} (1 + \alpha x^{-2})^{-(\alpha+1)/2} \, dx \\
&= \alpha^{(\alpha+1)/2} K_\alpha \int_t^\infty x^{-\alpha-1} - \frac{\alpha(\alpha+1)}{2} x^{-\alpha-3} + O(x^{-\alpha-5}) \, dx \\
&= \alpha^{(\alpha+1)/2} K_\alpha \left(\frac{1}{\alpha} t^{-\alpha} - \frac{\alpha(\alpha+1)}{2(\alpha+2)} t^{-\alpha-2} + O(t^{-\alpha-4})\right).
\end{aligned}$$

Consequently,

$$\overline{F}(\lambda t) - \lambda^{-\alpha}\overline{F}(t) \sim \alpha^{(\alpha+1)/2} K_\alpha \frac{\alpha(\alpha+1)}{2(\alpha+2)}(-\lambda^{-\alpha-2} + \lambda^{-\alpha})t^{-\alpha-2}$$

$$\sim \lambda^{-\alpha}\frac{\alpha^2(\alpha+1)}{2(\alpha+2)}\overline{F}(t)t^{-2}(1-\lambda^{-2}).$$

Therefore, $\overline{F}$ belongs to $2RV(-\alpha, \mathrm{Id}^{-2})$.

A first-order analysis shows that $G_*^{\leftarrow}(1-u) \asymp u^{1/\alpha}$. Consequently, choosing $k_n = n^{4\theta/(4+\alpha)}$ with θ positive less than 1 ensures that the condition listed in Proposition 4.2.2.v is satisfied. For such choice, the second assumption in (4.2.8) holds. We conclude that the distributional assumptions in the Resnick and Stărică theorem are satisfied.

An example of a process where the theorem leads to a fully explicit result is the AR(1) model, $X_n = \sum_{j \geq 0} r^j Z_{n-j}$ with $|r|$ less than 1. In that case

$$\lambda = \frac{1+|r|^\alpha}{\alpha^2(1-|r|^\alpha)}.$$

Note that when the order of the autoregressive process is known, Resnick and Stărică (1997) show that estimating the tail index by Hill's estimator applied to the estimated residuals yields a more efficient procedure. In contrast, the Hill estimator based on the observations is insensitive to misspecification of the the order of the model within the class of linear processes.

4.3. Randomly weighted sums.

In this subsection, we consider a weighted sum $\sum_{i \in \mathbb{Z}} w_i X_i$, where the weights $W = (w_i)_{i \in \mathbb{Z}}$ are random, independent of the X_i's. We also write $\langle W, X \rangle$ for this series. Clearly, under some assumptions on the weights, the uniformity of Theorem 2.5.1 allows one to obtain an asymptotic expansion for the tail of the weighted sum given the weights, and then decondition. This can be achieved with various integrability hypotheses on W, according to the arguments used in the proof. The one which we provide seems to work well for the applications which we will study. In particular, it does not add any moment requirement to the distribution of the X_i's. In applications, it is often assumed that the weights are nonnegative. This is not strictly necessary for deriving tail expansions, but it somewhat simplifies the statements and proofs.

We will develop some applications in the next subsections.

Before stating our main result on randomly weighted sums, recall that $\|\cdot\|_p$ is the ℓ_p-norm on sequences. Hence, when the sequence W is nonnegative, W_p is $\sum_{i \in \mathbb{Z}} w_i^p$. This is always defined, possibly infinite.

THEOREM 4.3.1. *Let F be a continuous distribution function, with tail $\overline{F}$ in $SR_{-\alpha,\omega}$ and such that $\overline{M_{-1}F} = O(\overline{F})$ at infinity. Let m be an integer less than α and ω. Let γ be a positive number less than 1 and $\omega - m$. Assume that*

$\overline{F}^{(m)}$ is bounded. Let $X = (X_i)_{i \in \mathbb{Z}}$ be a sequence of independent and identically distributed random variables having distribution F. Consider nonnegative random weights $W = (w_i)_{i \in \mathbb{Z}}$, independent of X, and such that for any $1 \leq k \leq j \leq m$, for some ϵ positive,

$$EW_1^{j-k} W_k (W_{\alpha+\epsilon} + W_{\alpha-\epsilon}) \mathbb{1}\{ N_{\alpha,\gamma,\omega}(W) > t \} = o(t^{-m}) \tag{4.3.1}$$

and

$$EW_1^{j-k} W_k \sum_{i \in \mathbb{Z}} w_i^{-j} \mathbb{1}\{ w_i > t \} = o\big(t^{-m}\overline{F}(t)\big) \tag{4.3.2}$$

as t tends to infinity, and

$$EN_{\alpha,\gamma,\omega}(W)^{m+\alpha+\epsilon} < \infty . \tag{4.3.3}$$

Then,

$$P\{ \langle W, X \rangle \geq t \} = \sum_{i \in \mathbb{Z}} EL_{F_{W\setminus w_i}, m} \overline{M_{w_i} F}(t) + o\big(t^{-m}\overline{F}(t)\big)$$

as t tends to infinity.

REMARK. The assumption that $\overline{F}^{(m)}$ is bounded is only used at the end of the proof to bound $w_i^{-j} \overline{F}^{(j)}(t/w_i)$ when t/w_i is of order 1, that is when w_i is large. If all the w_i's are less than some fixed number, boundedness of $F^{(m)}$ is not needed in the argument and assumption (4.3.2) holds. For example, the case of the w_i's being Bernoulli random variables occurs in the analysis of compound sums.

PROOF. Let R be $1/N_{\alpha,\gamma,\omega}(W)$. Define the sequence $C = RW$, whose elements are $c_i = w_i/N_{\alpha,\gamma,\omega}(W)$. Since $N_{\alpha,\gamma,\omega}(\cdot)$ is homogenous of degree 1, this new random sequence satisfies $N_{\alpha,\gamma,\omega}(C) = 1$. The conditional distribution function of $\langle C, X \rangle$ given C is F_C. Let ϵ be a positive real number. Let t_2 be as in Lemma 5.2.4 as applied to the normalized regularly varying function $\overline{F}$. Furthemore, let t_1 be at least t_2, and such that the function $\eta(\cdot)$ in Theorem 2.5.1 is at most ϵ on $[t_1, \infty)$. We apply Theorem 2.5.1 conditioning on W. So, on the event $\{ Rt > t_1 \}$,

$$\left| \overline{F}_C(Rt) - \sum_{i \in \mathbb{Z}} L_{F_{C\setminus c_i}, m} \overline{M_{c_i} F}(Rt) \right| \leq (Rt)^{-m} \overline{F}(Rt)\epsilon .$$

Let K be the (unconditional) distribution function of $\langle W, X \rangle$. Clearly, $K(t)$ is the expected value of $F_C(Rt)$. Taking expectation with respect to the sequence W in the previous inequality,

$$\left| \overline{K}(t) - \sum_{i \in \mathbb{Z}} EL_{F_{C\setminus c_i}, m} \overline{M_{c_i} F}(Rt) \right|$$
$$\leq \epsilon t^{-m} ER^{-m}\overline{F}(Rt) \mathbb{1}\{ Rt \geq t_1 \} + E\overline{F}_C(Rt)\mathbb{1}\{ Rt \leq t_1 \} \tag{4.3.4}$$
$$+ \sum_{i \in \mathbb{Z}} E| L_{F_{C\setminus c_i}, m} \overline{M_{c_i} F}(Rt) | \mathbb{1}\{ Rt \leq t_1 \} .$$

The equality
$$L_{F_{C\setminus c_i},m}\overline{M_{c_i}F} = L_{M_R F_{W\setminus w_i},m}M_R\overline{M_{w_i}F}$$
and Lemma 5.1.4 show that the left hand side of (4.3.4) is the absolute value of $\overline{K}(t)$ minus the asymptotic expansion given in the statement of the theorem.

To bound the right hand side of (4.3.4), the Potter bound in Lemma 5.2.4 implies that for Rt and t at least t_1,
$$\overline{F}(Rt)/\overline{F}(t) \leqslant R^{-\alpha}(R^\epsilon \vee R^{-\epsilon}).$$

Consequently,
$$ER^{-m}\overline{F}(Rt)\mathbb{1}\{Rt \geqslant t_1\} \leqslant ER^{-m-\alpha}(R^\epsilon \vee R^{-\epsilon})\overline{F}(t).$$

Next, we have the obvious inequality
$$EF_C(Rt)\mathbb{1}\{Rt \leqslant t_1\} \leqslant P\{Rt \leqslant t_1\} = P\{N_{\alpha,\gamma,\omega}(W) \geqslant t/t_1\}.$$

Using our integrability assumption on the weights and Markov's inequality, this is at most $o\bigl(t^{-m}\overline{F}(t)\bigr)$.

We now bound the third term in the right hand side of (4.3.4). Let F_* be the distribution function of $|X_i|$ and let H_C be the conditional one of $\sum_{i\in\mathbb{Z}}c_i|X_i|$ given C. Then, $\mu_{F_{C\setminus c_i},j}$ is at most $\mu_{H_C,j}$. Consequently,
$$E|L_{F_{C\setminus c_i},m}\overline{M_{c_i}F}(Rt)|\,\mathbb{1}\{Rt\leqslant t_1\}$$
$$\leqslant \sum_{0\leqslant j\leqslant m}E\mu_{H_C,j}|\mathrm{D}^j\overline{M_{c_i}F}(Rt)|\,\mathbb{1}\{Rt\leqslant t_1\}. \qquad (4.3.5)$$

We then make use of the following claim which allows us to untangle the weights and the random variables X. It is a special case of a lemma in Chow and Teicher (1978, §10.3). It controls the moments of a deterministically weighted sum by that of the random variables and various ℓ_p-norms of the weights. Recall the notation $C_k = \sum_{i\in\mathbb{Z}}c_i^k$.

CLAIM. *Let $p_{k,j} = k - 1 + (j-k)(j-k+1)/2$. For any positive integer j less than α,*
$$\mu_{H_C,j} \leqslant \sum_{1\leqslant k\leqslant j} 2^{p_{k,j}}\mu_{F_*,k}\mu_{F_*,1}^{j-k}C_1^{j-k}C_k.$$

PROOF. The proof is that of Chow and Teicher (1978, §10.3) but with the constant made explicit. For any nonnegative sequence $(a_i)_{i\in\mathbb{Z}}$,
$$\left(\sum_{i\in\mathbb{Z}}a_i\right)^j = \sum_{i\in\mathbb{Z}}a_i\left(a_i + \sum_{k\in\mathbb{Z}\setminus\{i\}}a_k\right)^{j-1}$$
$$\leqslant 2^{j-1}\left(\sum_{i\in\mathbb{Z}}a_i^j + \sum_{i\in\mathbb{Z}}a_i\left(\sum_{k\in\mathbb{Z}\setminus\{i\}}a_k\right)^{j-1}\right).$$

Substituting $c_i|X_i|$ for a_i in this inequality and taking expectation on both sides with respect to X,

$$\mu_{H_c,j} \leqslant 2^{j-1}\left(\mu_{F_*,j}C_j + \mu_{F_*,1}\sum_{i\in\mathbb{Z}} c_i E\left(\sum_{k\in\mathbb{Z}\setminus\{i\}} c_k X_k\right)^{j-1}\right)$$

$$\leqslant 2^{j-1}\left(\mu_{F_*,j}C_j + \mu_{F_*,1}\sum_{i\in\mathbb{Z}} c_i \mu_{H_c,j-1}\right)$$

$$= 2^{j-1}(\mu_{F_*,j}C_j + \mu_{F_*,1}C_1\mu_{H_c,j-1}).$$

The claim follows by induction. ∎

Continuing the proof of Theorem 4.3.1, that is, to bound (4.3.5) from above, we have,

$$\mathrm{D}^j \overline{M_{c_i}F}(Rt) = R^{-j}w_i^{-j}\overline{F}^{(j)}(t/w_i).$$

Recall that $\overline{F}^{(j)} \sim (-\alpha)_j \mathrm{Id}^{-j}\overline{F}$. Let ϵ be a positive real number. Potter's bounds ensure the existence of t_0 such that if both t and t/w_i are at least t_0, then

$$w_i^{-j}\overline{F}^{(j)}(t/w_i) \leqslant 2(-\alpha)_j(w_i^{\alpha+\epsilon} \vee w_i^{\alpha-\epsilon})t^{-j}\overline{F}(t).$$

Moreover, since $\overline{F}^{(m)}$ is bounded by assumption, so are all the j-th derivatives of F whenever j is at most m. For the following, it is convenient to introduce the quantity $C_0 = 1$. Using our claim, there exists a constant M, independent of i, such that each summand in (4.3.5) is at most

$$MEC_1^{j-k}C_k R^{-j}(w_i^{\alpha+\epsilon} + w_i^{\alpha-\epsilon})\mathbb{1}\{Rt \leqslant t_1\}\overline{F}(t)$$
$$+ MEC_1^{j-k}C_k R^{-j}w_i^{-j}\mathbb{1}\{w_i > t/t_0\}.$$

Since $C_s = W_s/N_{\alpha,\gamma,\omega}(W)^s$, the third sum in the right hand side of (4.3.4) is at most the sum for j nonnegative and at most m of

$$O(1)EW_1^{j-k}W_k(W_{\alpha+\epsilon} + W_{\alpha-\epsilon})\mathbb{1}\{N_{\alpha,\gamma,\omega}(W) > t/t_1\}\overline{F}(t)$$
$$+ O(1)EW_1^{j-k}W_k\sum_{i\in\mathbb{Z}} w_i^{-j}\mathbb{1}\{w_i > t/t_0\}.$$

Our integrability conditions imply that this last expression is $o(t^{-m}\overline{F}(t))$. ∎

4.4. Compound sums.

Theorem 4.3.1 has many applications in applied probability. In this subsection, we obtain a tail expansion for compound sums. Compound sums are a basic model in insurance mathematics, e.g. in modelling total claim size. A discussion of their asymptotic behavior may be found in Embrechts, Klüppelberg and Mikosch (1997). Some practical methods for obtaining tail area approximations for compound distributions may be found in Beirlant, Teugels and Vynckier (1996). Willmot and Lin (2000) is a good source of information for compound distributions and may be consulted for additional references; see also Willekens (1989). We further mention Omey and Willekens (1986, 1987) who obtained second-order results. We also mention Geluk (1992, 1996) who provides second-order results for subordinated probability distributions in the heavy tail case.

Let N be a nonnegative random variable, independent of the X_i's. Define the sum $S_n = X_1 + \cdots + X_n$, with $S_0 = 0$. Consequently, we agree that $F^{\star 0}$ is the point mass at 0. We write K for the distribution function of S_N.

THEOREM 4.4.1. *Assume that F is a distribution function on the nonnegative half-line, such that $\overline{F}$ belongs to $SR_{-\alpha,\omega}$. Let m be an integer less than α and ω. Let γ be a positive number less than $\omega - m$ and 1. If N has a moment of order more than*
$$\frac{\alpha+m}{\gamma}\left(\frac{\alpha+\omega}{\alpha} + 2 + \gamma\right),$$
then,
$$\overline{K} = EN L_{F^{\star(N-1)},m}\overline{F} + o(\mathrm{Id}^{-m}\overline{F})$$
at infinity.

PROOF. Let W be the random sequence defined by $w_i = \mathbb{1}\{0 < i \leqslant N\}$. Then $S_N = \langle W, X \rangle$. To apply Theorem 4.3.1 with the remark following it, we need to check its assumptions, which is to check the integrability condition on the weights. Since
$$\frac{1}{\gamma}\left(\frac{\alpha}{\alpha+\omega} \wedge \frac{1}{2}\right)^{-1} = \frac{1}{\gamma}\left(\frac{\alpha+\omega}{\alpha} \vee 2\right),$$
we have
$$N_{\alpha,\gamma,\omega}(W) = N^{\frac{1}{\gamma}(\frac{\alpha+\omega}{\alpha} \vee 2)} \vee 2^{\frac{\alpha}{\alpha+\omega}}.$$
Moreover, $W_1 = N$ and $W_k = N$. So the integrability conditions are implied by
$$EN^{\frac{m+\alpha+\epsilon}{\gamma}(\frac{\alpha+\omega}{\alpha} \vee 2)} < \infty,$$
and
$$EN^{m+\alpha+\epsilon}\mathbb{1}\{N > t^{\gamma(\frac{\alpha}{\alpha+\omega} \wedge \frac{1}{2})}\} = o(t^{-m}).$$
The first expectation is finite by assumption. For the second one, we apply the standard trick to prove Markov's inequality. For any p, the expectation is at most
$$EN^{m+\alpha+p+\epsilon} t^{-p\gamma(\frac{\alpha}{\alpha+\omega} \wedge \frac{1}{2})}.$$
Take p such that the exponent of t is less than $-m$ but such that $EN^{m+\alpha+p+\epsilon}$ is finite. Use Proposition 1.3.6 in Bingham, Goldie and Teugels (1989) to conclude that the assumptions of Theorem 4.3.1 hold. To conclude, note that for i positive and less than N, the equalities $M_{w_i}F = F$ and $F_{W \setminus w_i} = F^{\star(N-1)}$ hold. ∎

As in Barbe and McCormick (2004), using the Laplace transforms of X_1 and N allows one to derive a rather neat expression. Indeed, setting
$$\Lambda_X(t) = Ee^{-tX} \quad \text{and} \quad \Lambda_N(t) = Ee^{-tN},$$

equality (2.2.1) in Barbe and McCormick (2004) yields

$$ENL_{F*(N-1),m} = -\sum_{0 \leqslant j \leqslant m} \frac{1}{j!} \frac{d^j}{du^j} \frac{\Lambda'_N(-\log \Lambda_X(u))}{\Lambda_X(u)} \bigg|_{u=0} D^j. \qquad (4.4.1)$$

Then, the technique explained in Barbe and McCormick (2004) allows for efficient computation using computer algebra packages.

Theorem 4.4.1 has an interesting special case.

COROLLARY 4.4.2. *Let F be a distribution function satisfying the assumptions of Theorem 4.4.1. If N has a Poisson distribution with parameter a, then*

$$\overline{K} = aL_{K,m}\overline{F} + o(\mathrm{Id}^{-m}\overline{F}).$$

PROOF. The equality

$$\sum_{k>0} kS^j_{k-1} \frac{a^k}{k!} e^{-a} = a \sum_{k \geqslant 1} S^j_{k-1} \frac{a^{k-1}}{(k-1)!} e^{-a}$$

shows that

$$ENL_{F*(N-1),m} = aEL_{F*N,m} = aL_{K,m}.$$

The result follows from Theorem 4.4.1. ∎

The expression in Corollary 4.4.2 involves the Laplace character of K and requires calculating the moments of K. The following result involves only F; in the statement, the exponential is taken in the ring $\mathbb{R}_m[D]$, and is defined through the usual series expansion.

COROLLARY 4.4.3. *With the notations and assumptions of Corollary 4.4.2,*

$$\overline{K} = ae^{a(L_{F,m}-\mathrm{Id})}\overline{F} + o(\mathrm{Id}^{-m}\overline{F}).$$

PROOF. In the ring $\mathbb{R}_m[D]$,

$$ENL^{N-1}_{F,m} = e^{-a} \sum_{n \geqslant 1} nL^{n-1}_{F,m} \frac{a^n}{n!} = e^{-a} a e^{aL_{F,m}} = ae^{a(L_{F,m}-\mathrm{Id})}.$$

The result follows from Theorem 4.4.1. ∎

The considerable advantage of the formula in Corollary 4.4.3 is evident in the following Maple (version 9) implementation,

```
m:=3:
mu[0]:=1:
LF:=sum('(-1)^j*mu[j]*D^j/j!','j'=0..infinity):
taylor(a*exp(a*(LF-1)),D=0,m+1);
```

which yields the first four terms,

$$E(NL_{F,3}^{N-1}) = a\mathrm{Id} - a^2\mu_{F,1}\mathrm{D} + \frac{a^2}{2}(a\mu_{F,1}^2 + \mu_{F,2})\mathrm{D}^2$$
$$- \frac{a^2}{6}(a^3\mu_{F,1}^2 + 3a\mu_{F,1}\mu_{F,2} + \mu_{F,3})\mathrm{D}^3.$$

Combined with Corollary 4.4.3, this shows that under the assumptions of Theorem 4.4.1 with m equal to 3, we have the four terms expansion

$$\overline{K} = a\overline{F} - a^2\mu_{F,1}\overline{F}' + \frac{a^2}{2}(a\mu_{F,1}^2 + \mu_{F,2})\overline{F}''$$
$$- \frac{a^2}{6}(a^3\mu_{F,1}^2 + 3a\mu_{F,1}\mu_{F,2} + \mu_{F,3})\overline{F}''' + o(\mathrm{Id}^{-3}\overline{F}).$$

4.5. Queueing theory.

We consider a queue of M/G/1 type. This means that customers arrive with interarrival time exponentially distributed with mean μ; the service has general distribution function B with finite mean β. We assume in this application that the system is stable, that is β/μ is less than 1. Define

$$H(t) = \beta^{-1}\int_0^t \overline{B}(s)\,\mathrm{d}s, \quad t \geqslant 0.$$

Set $a = \beta/\mu$. The Pollaczek-Hincin formula shows that the waiting time has the compound geometric distribution function

$$W = (1-a)\sum_{n\geqslant 0} a^n H^{\star n}.$$

We refer to Bingham, Goldie and Teugels (1987, p. 387) or Cohen (1972) for this derivation. If $\overline{B}$ is regularly varying at infinity, Cohen (1972) shows that so are $\overline{H}$ and $\overline{W}$. When $\overline{B}$ satisfies the assumption of Theorem 2.5.1, higher-order expansions can be obtained. We mention Abate, Choudhury and Whitt (1994), Abate, Choudhury, Lucantoni and Whitt (1995), Abate and Whitt (1997) for recent related work. See also Willekens and Teugels (1992) and Whitt (2002) where further references may be found.

PROPOSITION 4.5.1. *Assume that B satisfies the assumptions of Theorem 4.4.1. Let Λ_H be the Laplace transform of H. Then*

$$\overline{W} = a(1-a)\sum_{0\leqslant j\leqslant m} \frac{1}{j!}\frac{\mathrm{d}^j}{\mathrm{d}u^j}\bigl(1 - a\Lambda_H(u)\bigr)^{-2}\Big|_{u=0}\overline{H}^j + o(\mathrm{Id}^{-m}\overline{H})$$

at infinity.

PROOF. Let S_n be a sum of n independent and identically distributed random variables having distribution function H. We set $S_0 = 0$. We see that W is the distribution of S_N where N has a geometric distribution with parameter a. The Laplace transform of N is

$$\Lambda_N(u) = (1-a)/(1-ae^{-u}).$$

Therefore, for a random variable X having distribution H,

$$\frac{\Lambda'_N(-\log \Lambda_X(u))}{\Lambda_X(u)} = \frac{-a(1-a)}{(1-a\Lambda_X(u))^2}.$$

Applying formula (4.4.1), we obtain the result. ∎

As with Corollary 4.4.2, an alternative expression is useful for computing.

COROLLARY 4.5.2. *With the notations and assumptions of Proposition 4.5.1,*

$$\overline{W} = a(1-a)(\mathrm{Id} - a\mathsf{L}_{H,m})^{-2}\overline{H} + o(\mathrm{Id}^{-m}\overline{H}).$$

PROOF. In the ring $\mathbb{R}_m[\mathrm{D}]$,

$$EN\mathsf{L}_{H,m}^{N-1} = (1-a)\sum_{n\geqslant 1} na^n \mathsf{L}_{H,m}^{n-1} = a(1-a)(\mathrm{Id} - a\mathsf{L}_{H,m})^{-2}.$$

The result follows from Theorem 4.4.1. ∎

The following Maple (version 9) code gives the first four terms of the expansion,

```
m:=3:
mu[0]:=1:
LF:=sum('(-1)^j*mu[j]*D^j/j!','j'=0..infinity):
simplify(taylor(a*(1-a)*(1-a*LF)^{-2}),D=0,m+1);
```

that is, with $b = 1 - a$,

$$EN\mathsf{L}_{F,m}^{N-1} = b\,\mathrm{Id} - 2b^2\mu_{F,1}\mathrm{D} + b^2(\mu_{F,2} + 3\mu_{F,1}^2)\mathrm{D}^2$$
$$+ \frac{b^3}{3}(-\mu_{F,1}^3 + 9b\mu_{F,1}\mu_{F,2} - 12b^2\mu_{F,1}^2)\mathrm{D}^3.$$

To be explicit, this formula and Corollary 4.5.2 provide the following four terms expansion under the assumptions of Proposition 4.5.1,

$$\overline{W} = b\overline{H} - 2b^2\mu_{F,1}\overline{H}' + b^2(\mu_{F,2} + 3\mu_{F,1}^2)\overline{H}''$$
$$+ \frac{b^3}{3}(-\mu_{F,1}^3 + 9b\mu_{F,1}\mu_{F,2} - 12b^2\mu_{F,1}^2)\overline{H}''' + o(\mathrm{Id}^{-3}\overline{H}).$$

4.6. Branching processes.

Consider an age dependent branching process. Basically, this refers to a Galton-Watson branching process in which the particles have a random life time governed by a distribution F. The process starts at time 0 with one particle; at the end of its life, it generates an average of ρ particles, which themselves at death will each generate an average of ρ particle, and so on. We refer to Athreya and Ney (1972, chapter 4) for a complete description of the process. It is intuitively clear that if ρ is less than 1, the so-called subcritical case, then the process will become extinct.

Let $\nu(t)$ be the expected number of particles alive at time t.

PROPOSITION 4.6.1. *Assume that F is a continuous distribution function on the positive half-line, whose tail $\overline{F}$ belongs to $SR_{-\alpha,\omega}$. Let m be an integer less than α and ω. Then*

$$\nu = (1-\rho)(\mathrm{Id} - \rho L_{F,m})^{-2}\overline{F} + o(\mathrm{Id}^{-m}\overline{F})$$

at infinity.

PROOF. As before, let us agree that $F^{\star 0}$ is the distribution function of the point mass at 0. Equation (4) in section IV.5 of Athreya and Ney (1972) shows that

$$\nu = \sum_{k\geqslant 0}\rho^k \overline{F^{\star(k+1)}} - \sum_{k\geqslant 0}\rho^k \overline{F^{\star k}}.$$

Let N be a geometric random variable with parameter ρ. Applying Theorem 4.4.1 under the form of Corollary 4.5.2, we see that

$$\sum_{k\geqslant 0}\rho^k \overline{F^{\star(k+1)}} = \rho^{-1}(1-\rho)^{-1}E(\overline{F^{\star N}}\mathbb{1}\{N\geqslant 1\})$$
$$= (\mathrm{Id} - \rho L_{F,m})^{-2}\overline{F} + o(\mathrm{Id}^{-m}\overline{F}).$$

Furthermore, still applying Theorem 4.4.1 under the form of Corollary 4.5.2,

$$\sum_{k\geqslant 0}\rho^k \overline{F^{\star k}} = \rho(\mathrm{Id} - \rho L_{F,m})^{-2}\overline{F} + o(\mathrm{Id}^{-m}\overline{F}).$$

The result follows. ∎

Let us now present an explicit expansion with $m = 2$. We write σ_F^2 for the variance of the distribution F. Direct calculation yields

$$(1-\rho)(\mathrm{Id} - \rho L_{F^{\star(N-1)},2})^{-2}$$
$$= \frac{1}{1-\rho}\mathrm{Id} - \frac{2\rho\mu_{F,1}}{(1-\rho)^2}\mathrm{D} + \frac{\rho}{(1-\rho)^3}\big((1-\rho)\sigma_F^2 + (1+2\rho)\mu_{F,1}^2\big)\mathrm{D}^2,$$

so that

$$\nu = \frac{1}{1-\rho}\overline{F} + \frac{2\rho\mu_F}{(1-\rho)^2}\overline{F}' - \frac{\rho}{(1-\rho)^3}\big((1-\rho)\sigma_F^2 + (1+2\rho)\mu_{F,1}^2\big)\overline{F}'' + o(\mathrm{Id}^{-2}\overline{F}).$$

The first-order term yields the result in Chover, Ney and Waigner (1973, p. 296) and the first two terms yield the second-order result result in Grubel (1987). We refer to Pakes (1975) for related work in a subexponential setting.

Note that if the Laplace transform of F is known, one could use Proposition 4.5.1 to obtain an alternative form of the asymptotic expansion.

4.7. Infinitely divisible distributions.

The infinite divisible distributions are those which for any integer n can be written as n-th convolution power of another distribution. They are also the limiting distributions of sums of n independent and identically distributed random variables when the distribution is allowed to change with n. We refer to Feller (1971) for an introduction to the topic. These distributions are characterized through their characteristic functions, which are of the form

$$\zeta \in \mathbb{R} \mapsto \exp\Big(i\tau\zeta - \frac{\sigma^2}{2}\zeta^2 + \int \Big(e^{i\zeta x} - 1 - i\frac{\zeta x}{1+x^2}\Big)\,\mathrm{d}\nu(x)\Big),$$

where σ^2 is nonnegative, τ is a real number, ν, the so-called Lévy measure, has no mass at the origin and satifies $\int x^2/(1+x^2)\,\mathrm{d}\nu(x) < \infty$. Note that throughout this subsection, i denotes the imaginary unit; this is the only part in this paper where complex numbers are used.

For what follows, it is convenient to set

$$\overline{\nu}(t) = \nu(t, \infty).$$

Let G_ν be an infinitely divisible distribution function with Lévy measure ν. If ν has a regularly varying tail, then $\overline{G}_\nu \sim \overline{\nu}$; see e.g. Feller (1971); see also Pakes (2004) for a related first-order result and Embrechts, Goldie and Veraverbeke (1979) for work in the subexponential case. Under a slightly stronger assumption, Grübel (1987) proved the two terms expansion

$$\overline{G}_\nu \sim \overline{\nu} - \mu_{G,1}\mathrm{D}\overline{\nu}$$

at infinity. With our notation this means $\overline{G}_\nu \sim \mathsf{L}_{G_\nu,1}\overline{\nu}$.

The following result is then quite natural.

PROPOSITION 4.7.1. *Let ν be a measure such that $\overline{\nu}$ is smoothly varying of index $-\alpha$ and order ω and $\nu(-\infty, -t) = O(\overline{\nu}(t))$ as t tends to infinity. For any integer m less than α and ω,*

$$\overline{G}_\nu = \mathsf{L}_{G_\nu,m}\overline{\nu} + o(\mathrm{Id}^{-m}\overline{\nu}).$$

In this expansion, the Laplace character $\mathsf{L}_{G_\nu,m}$ involves the moments of G_ν of order at most m. These are finite when $\overline{\nu}$ is regularly varying of index less than $-m$ and the tails of ν are either balanced or the right tail is dominant, that is under the assumptions of the Proposition. In practice, those moments can be computed by differentiating the characteristic function of G_ν. Again, computer algebra packages are wonderful for doing this type of work.

PROOF. The basic idea underlying the proof is a classical representation of infinitely divisible distributions as convolutions of two distributions, the first one having light tail, the second one being that of a randomly stopped sum with heavy tail. This makes it possible to use Theorem 4.4.1.

Let A be a positive constant, and let a be the ν-measure of $[-A, A]^c$. Let ν_1 be the measure $\nu(\cdot \cap [-A, A]^c)$, and let F be the distribution function pertaining to the probability measure ν_1/a. Define

$$\tau_1 = \tau - \int_{[-A,A]^c} \frac{x}{1+x^2} \, d\nu(x).$$

The function

$$\widehat{H}(\zeta) = \exp\left(i\zeta\tau_1 - \frac{\sigma^2}{2}\zeta^2 + \int_{-A}^{A} e^{i\zeta x} - 1 - \frac{i\zeta x}{1+x^2} \, d\nu(x)\right)$$

is the characteristic function of an infinitely divisible distribution function H. Write $\widehat{F}$ for the characteristic function of F and $\widehat{G_\nu}$ for that of G_ν. One has

$$\widehat{G_\nu} = \widehat{H} \exp\left(a(\widehat{F} - 1)\right). \tag{4.7.1}$$

Let $X = (X_i)_{i \geqslant 1}$ be a sequence of independent and identically distributed random variables, all having distribution F. They induce the partial sums $S_n = X_1 + \cdots + X_n$, with the usual convention $S_0 = 0$. Let N be a Poisson random variable, with mean a. Let K be the distribution function of the randomly stopped sum S_N. One can check that the characteristic function of K is

$$\widehat{K} = \exp\left(a(\widehat{F} - 1)\right),$$

so that $\widehat{G_\nu} = \widehat{H}\widehat{K}$. Consequently, $G_\nu = H \star K$.

It follows from Corollary 4.4.2 that $\overline{K} \sim a L_{K,m} \overline{F}$.

The remainder of the proof is the only place in section 4 where we rely on results which we will establish in the proof of Theorem 2.5.1. We write

$$\overline{G}_\nu(t) = \int_{-\infty}^{t/2} \overline{K}(t-x) \, dH(x) + \int_{-\infty}^{t/2} \overline{H}(t-x) \, dK(x) + \overline{H}\,\overline{K}(t/2).$$

As observed by Grübel (1987, proof of Theorem 7), the function $\overline{H}$ decays exponentially fast to 0 at infinity. So

$$\overline{G}_\nu(t) = \int_{-\infty}^{t/2} \overline{K}(t-x) \, dH(x) + o\left(t^{-m}\overline{F}(t)\right).$$

It follows from Theorem 6.4.1, Lemma 7.2.1 and the asymptotic expansion of $\overline{K}$ that

$$\overline{G}_\nu(t) = \sum_{0 \leqslant j \leqslant m} \frac{(-1)^j}{j!} a\mu_{K,j} \int_{-\infty}^{t/2} \mathrm{D}^j \overline{F}(t-x) \,\mathrm{d}H(x) + o\bigl(t^{-m}\overline{F}(t)\bigr)$$

$$= \sum_{0 \leqslant j \leqslant m} \frac{(-1)^j}{j!} a\mu_{K,j} L_{H,m-j} \mathrm{D}^j \overline{F}(t) + o\bigl(t^{-m}\overline{F}(t)\bigr).$$

Applying Lemma 5.1.3, we conclude that

$$\overline{G}_\nu = a L_{K \star H, m} \overline{F} + o(\mathrm{Id}^{-m}\overline{F}).$$

Since $K \star H$ is G_ν and $a\overline{F}$ ultimately coincide with $\overline{\nu}$, this concludes the proof. ∎

4.8. Implicit transient renewal equation and iterative systems.

A renewal equation is an integral equation of the form

$$F - aF \star H = K,$$

where H and K are given distribution functions, a is a real number and F is the unknown. Such equation is transient if the absolute value of a is less than 1. These equations arise in applied probability and we refer to Feller (1971) for an introduction to renewal theory and Bingham, Goldie and Teugels (1989, §8.6) for basic results and references on the tail behaviour of the solutions. Following Goldie (1991), implicit renewal equations are equations of the same form, except that K is an integral transform of F. Again, these equations appear in applied probability and statistics, in connection with the stationary solution of iterative systems.

The purpose of this subsection is to show that in some cases, one can derive an asymptotic expansion of the solution by solving linear equations. This method is radically different from that of Kesten (1973) or Goldie (1991); it works under a different set of assumptions, closer to the one used by Grey (1994). The appealing feature of this approach is its simplicity of implementation. The idea is very elementary: expand the known function in an asymptotic scale; in this scale, the implicit renewal equation becomes a finite dimensional linear system, which can be solved by elementary linear algebraic techniques. Consequently, we obtain an analogue of the technique to solve differential equations using formal series expansions and identifying the coefficients. Of course, this technique will only work for a limited type of renewal equation.

To explain the principle, we first look at a simple renewal equation.

Standard renewal equation. Assume that H and K have moments of order m. Then the renewal equation gives m equations determining the moments of F, that is

$$\mu_{F,j} - a \sum_{0 \leqslant i \leqslant j} \binom{j}{i} \mu_{F,i} \mu_{H,j-i} = \mu_{K,j}, \qquad 1 \leqslant j \leqslant m.$$

4.8. IMPLICIT TRANSIENT RENEWAL EQUATION AND ITERATIVE SYSTEMS

This can be rephrased in a nicer form with Proposition 2.3.1, which combined with the renewal equation imply

$$L_{F,m} - aL_{F,m}L_{H,m} = L_{K,m}.$$

When the absolute value of a is less than 1, the operator $\mathrm{Id} - aL_{H,m}$ is not in the ideal generated by D. Therefore, it is invertible in $\mathbb{R}_m[\mathrm{D}]$. This implies that for $|a|$ less than 1,

$$L_{F,m} = (\mathrm{Id} - aL_{H,m})^{-1} L_{K,m}. \tag{4.8.1}$$

So, we can calculate the Laplace character and the moments of F. This calculation is done by manipulating finite dimensional matices.

Iterating the renewal equation yields

$$F = \sum_{i \geq 0} a^i K \star H^{\star i}. \tag{4.8.2}$$

Hence, F converges to $1/(1-a)$ at infinity and its tail is given by

$$(1-a)^{-1} - F = \sum_{i \geq 0} a^i \overline{K \star H^{\star i}}.$$

Define the distribution function G by $\overline{G} = 1 - (1-a)F$. It solves the equation

$$\overline{G} - a\overline{G} \star H = (1-a)\overline{K}. \tag{4.8.3}$$

Similarly to what we did in subsection 3.3, assume that K and H have an asymptotic expansion in a $\star$-asymptotic scale e. It is then conceivable that $\overline{G}$ has an expansion in that scale. Then (4.8.3) and Theorem 2.5.1 yield, with notation analogous to that of section 3,

$$p_{\overline{G}} - a(\mathcal{L}_G p_{\overline{H}} + \mathcal{L}_H p_{\overline{G}}) = (1-a)p_{\overline{K}}.$$

That is, $p_{\overline{G}}$ is the solution of a linear system of equations, and can be made explicit by the formula

$$p_{\overline{G}} = (\mathrm{Id} - a\mathcal{L}_H)^{-1}\big((1-a)p_{\overline{K}} + a\mathcal{L}_G p_{\overline{H}}\big). \tag{4.8.4}$$

In this formula, $\mathcal{L}_G$ involves the moments of G and those are $1 - a$ times the moments of F; these can be calculated by solving the linear system (4.8.1). Having an expansion for $\overline{G}$ it is then trivial to derive one for F.

The only thing missing to make this rigorous is to prove that G has an asymptotic expansion in the same asymptotic scale e as H and K. This is trivial because, when a is positive, representation (4.8.2) shows that G is the distribution function of the sum of a random variable with distribution K and a compound sum having a number of summands distributed according to a geometric distribution with parameter a; and then Theorem 4.4.1 applies. When a is negative, one uses the same argument but split the sum in (4.8.2) into one where the index i is odd and one where the index is even.

An implicit renewal equation. Following Goldie (1991), Grey (1994), Grincevičius (1975) and Kesten (1973), consider the distributional equation involving random variables, $R \stackrel{d}{=} Q + MR$ where (M,Q) and R are independent and all random variables are nonnegative. We need one important extra assumption, namely that M and Q are independent. It would be desirable not to make this assumption, but this would require some interesting generalization of Theorem 2.5.1. To write the implicit renewal equation equivalent to this distributional identity, let $\stackrel{M}{\star}$ denote the Mellin-Stieltjes convolution. That is, if F and H are two distribution functions,

$$H \stackrel{M}{\star} F(t) = \int F(t/x) \, dH(x).$$

The connection with iterative systems is that if (Q_i, M_i), $i \geqslant 1$, are independent, all distributed as (Q, M), then, under suitable conditions, the sequence $R_0 = 0$, $R_n = Q_n + M_n R_{n-1}$ has a distribution converging to that of R, which we denote by F — see for instance the survey article by Diaconis and Freedman (1999). We mention that first-order asymptotics for random coefficient autoregressive models is considered in Resnick and Willekens (1991).

Let H and K be the distribution functions of M and Q respectively. The distributional equation is equivalent to

$$F = K \star (H \stackrel{M}{\star} F). \tag{4.8.5}$$

Let us assume that K has a tail expansion in a $\star$-asymptotic scale e as in section 3.3. Equation (4.8.5) yields two equations: one on the Laplace characters, which allows us to identify the moments of F, one on the tail vectors $p_{\overline{F}}$, $p_{\overline{K}}$, which allows one to find the tail expansion of the solution.

For any integer k for which the k-th moments of K and H exist, the moment equations are

$$\mu_{F,k} = \sum_{0 \leqslant j \leqslant k} \binom{k}{j} \mu_{K,j} \mu_{H \stackrel{M}{\star} F, k-j}$$

$$= \sum_{0 \leqslant j \leqslant k} \binom{k}{j} \mu_{K,j} \mu_{H, k-j} \mu_{F, k-j}.$$

These equations allow one to find the moments $\mu_{F,k}$ by induction on k. With the notation introduced in section 2.1, they are written in terms of Laplace characters as

$$L_{F,m} = L_{K,m}(L_{H,m} \stackrel{M}{\circ} L_{F,m}),$$

an expression suitable for implementation on a computer algebra package.

If $\overline{F}$ has an expansion in the scale e, then (4.8.5) and Theorem 2.5.1 suggest that

$$p_{\overline{F}} = \mathcal{L}_{H \stackrel{M}{\star} F} p_{\overline{K}} + \mathcal{L}_K p_{\overline{H \stackrel{M}{\star} F}}.$$

Since the variables are nonnegative,

$$\overline{H \stackrel{M}{\star} F}(t) = \int \overline{F}(t/x) \, dH(x) = \int \overline{M_x F}(t) \, dH(x).$$

Consequently,
$$p_{\overline{H^{\mathrm{M}}_\star F}} = \int \mathcal{M}_x \, dH(x) \, p_{\overline{F}}.$$

Thus, we obtain the tail expansion of F in the scale e, whose coefficients are given by
$$p_{\overline{F}} = \left(\mathrm{Id} - \mathcal{L}_K \int \mathcal{M}_x \, dH(x)\right)^{-1} \mathcal{L}_{H^{\mathrm{M}}_\star F} p_{\overline{K}}. \tag{4.8.6}$$

Again, for this to be justified, we only need to prove that $\overline{F}$ has an expansion in the scale e. Our next result gives a sufficient condition, and a complete example of application follows its proof.

PROPOSITION 4.8.1. *Assume that K satistifies the assumptions of Theorem 4.3.1. If $EM_1^{2(\alpha+m+1)}$ is less than 1, then formula (4.8.6) holds.*

Note that by Jensen's inequality, the integrability assumption in Proposition 4.8.1 implies that $E \log M_1$ is negative, and therefore, that the implicit renewal equation has a well defined solution.

PROOF. Let Q_i (respectively M_i), $i \geqslant 1$, be a sequence of independent random variables all having the distribution K (respectively H). Let $w_0 = 1$ and for k at least 1, let $w_k = M_1 \ldots M_k$. We write W for the nonnegative sequence $(w_i)_{i \in \mathbb{N}}$. Under the assumption of Proposition 4.8.1, R has the same distribution as $\sum_{i \geqslant 1} w_{i-1} Q_i$. Therefore, we will derive the proposition from Theorem 4.3.1. So, we need to check that (4.3.1), (4.3.2) and (4.3.3) hold. Referring to (4.3.1), we set $p = j - k$ and take ϵ less than $1/4$. Clearly $p + k$ is at most m. Moreover, define
$$\rho = \frac{1}{2} \wedge \frac{\alpha}{\alpha + \omega}, \quad \text{and} \quad q = \left\lfloor \frac{\alpha + m}{\gamma \rho} + 1 \right\rfloor.$$

Furthermore, set $s = \gamma \rho q$. Note that $t^{-s} = o(t^{-m} \overline{F}(t))$ at infinity, for s is more than $m + \alpha$.

Applying Hölder's inequality, we bound the expectation involved in (4.3.1) by a constant times
$$\left(EW_1^{2p} W_k^2 (|w|_{\alpha+\epsilon}^2 + W_{\alpha-\epsilon}^2)\right)^{1/2} \left(EN_{\alpha,\gamma,\omega}(W)^{2s}\right)^{1/2} t^{-s}. \tag{4.8.7}$$

Let η be either $\alpha + \epsilon$ or $\alpha - \epsilon$. We have
$$W_1^{2p} W_k^2 W_\eta^2 = (2p)! \, 2! \, 2! \sum_{n_1 < \cdots < n_{2p+4}} w_{n_1} \cdots w_{n_{2p}} w_{n_{2p+1}}^k w_{n_{2p+2}}^k w_{n_{2p+3}}^\eta w_{n_{2p+4}}^\eta$$
$$+ \text{other terms where } n_i = n_j \text{ for some distinct } i \text{ and } j. \tag{4.8.8}$$

In all the terms in the right hand side of (4.8.8), the M_i's are raised to a power at most $2(p + k + \eta)$, which is at most $2(m + \eta)$. Concerning the first sum on the right hand side,

$$
\begin{aligned}
Ew_{n_1}&\cdots w_{n_{2p}}w_{n_{2p+1}}^k w_{n_{2p+2}}^k w_{n_{2p+3}}^\eta w_{n_{2p+4}}^\eta \\
&= EW_{n_1}^{2(p+k+\eta)}\left(\frac{w_{n_2}}{w_{n_1}}\right)^{2(p+k+\eta)-1}\cdots\left(\frac{w_{n_{2p+1}}}{w_{n_{2p}}}\right)^{2(k+\eta)} \\
&\quad \times\left(\frac{w_{n_{2p+2}}}{w_{n_{2p+1}}}\right)^{k+2\eta}\left(\frac{w_{n_{2p+3}}}{w_{n_{2p+2}}}\right)^{2\eta}\left(\frac{w_{n_{2p+4}}}{w_{n_{2p+3}}}\right)^{\eta} \\
&= \mu_{H,2(p+k+\eta)}^{n_1}\mu_{H,2(p+k+\eta)-1}^{n_2-n_1}\cdots\mu_{H,2(k+\eta)}^{n_{2p+1}-n_{2p}} \\
&\quad \times \mu_{H,k+2\eta}^{n_{2p+2}-n_{2p+1}}\mu_{H,2\eta}^{n_{2p+3}-n_{2p+2}}\mu_{H,\eta}^{n_{2p+4}-n_{2p+3}}.
\end{aligned} \quad (4.8.9)
$$

By Lyapounov's inequality, for s at most $2(m+\eta)$, the inequality $\mu_{H,s}\leqslant \mu_{H,2(m+\eta)}^{s/2(m+\eta)}$ holds. Therefore, (4.8.9) is at most $\mu_{H,2(m+\eta)}$ at the power

$$
\frac{1}{2(m+\eta)}\Big(2(p+k+\eta)n_1 + (2(p+k+\eta)-1)(n_2-n_1)+\cdots
$$
$$
+2(k+\eta)(n_{2p+1}-n_{2p}) + (k+2\eta)(n_{2p+2}-n_{2p+1})
$$
$$
+2\eta(n_{2p+3}-n_{2p+2}) + \eta(n_{2p+4}-n_{2p+3})\Big).
$$

This exponent is at least $\eta n_{2p+4}/2(m+\eta)$. Our moment assumption and Lyapounov's inequality imply that $\mu_{H,2(m+\eta)}$ is less than 1. Consequently, (4.8.9) is at most $\mu_{H,2(m+\eta)}^{\eta n_{p+4}/2(m+\eta)}$. For n_{2p+4} fixed, there are at most n_{2p+4}^{2p+3} integers $n_1,\ldots,n_{2p+3}$ less than n_{2p+4}. Therefore, the expectation of the first sum in the right hand side of (4.8.8) is at most

$$
\sum_{n\geqslant 1} n^{2p+3}(\mu_{H,2(m+\eta)}^{k/2(m+\eta)})^n,
$$

which is finite.

The other sums in the right hand side of (4.8.8) are similarly shown to have finite expectation.

So, to check (4.3.1) it remains to prove that $N_{\alpha,\gamma,\omega}(W)^{2s}$ has finite expectation. It suffices to show that $E\|W\|_{\gamma\rho}^{2s}$ and $E\|W\|_{\infty}^{2s}$ are finite.

We have

$$
E\|W\|_{\gamma\rho}^{2s} = E\Big(\sum_{i\geqslant 1}w_i^{\gamma\rho}\Big)^{2q} = \sum_{i_1,\ldots,i_{2q}}w_{i_1}^{\gamma\rho}\cdots w_{i_{2q}}^{\gamma\rho}.
$$

Again, this sum involves M_i's raised at power at most $2q\gamma\rho$, that is $2s$, which is less than $2(\alpha+m+\gamma\rho)$, which is less than $2(\alpha+m+1)$. By assumption, $\mu_{H,2(\alpha+m+1)}$ is less than 1, and the same argument as before shows that $E\|W\|_{\gamma\rho}^{2s}$ is finite.

Finally, note that

$$
P\{\|W\|_\infty^{2s}\geqslant t\} \leqslant \sum_{i\geqslant 1} P\{w_i\geqslant t^{1/2s}\}
$$
$$
\leqslant \sum_{i\geqslant 1} t^{-(\alpha+m+1)/s}Ew_i^{2(\alpha+m+1)}
$$
$$
\leqslant t^{-(\alpha+m+1)/s}(1-\mu_{H,2(\alpha+m+1)})^{-1}.
$$

4.8. IMPLICIT TRANSIENT RENEWAL EQUATION AND ITERATIVE SYSTEMS

Since $(\alpha + m + 1)/s$ is more than 1, this shows that $\|W\|_\infty^{2s}$ has finite expectation.

To check (4.3.2), we take $\eta = \alpha + \epsilon$. The indicator function of the event $w_i > t$ is at most $w_i^{m+\eta}/t^{m+\eta}$. Furthermore, $t^{-m-\eta} = o(t^{-m}\overline{F}(t))$. Consequently (4.3.2) is implied by

$$EW_1^{j-k}W_k W_{m+\eta} < \infty.$$

The same arguments as used to check (4.3.1) establish the above finiteness condition.

Finally, it is now straigtfoward to check that (4.3.3) holds. ∎

Let us now give an explicit example. Its purpose is to show that it is now easy to obtain the tail expansion, at least when the assumptions of Proposition 4.8.1 are satisfied. So for our example, take H to be the exponential distribution function of mean θ,

$$H(t) = 1 - e^{-t/\theta}, \qquad t \geqslant 0,$$

and K to be the Pareto distribution

$$K(t) = 1 - (1+t)^{-\alpha}, \qquad t \geqslant 0.$$

To check the assumptions of Proposition 4.8.1, we calculate

$$EM^\lambda = \int x^\lambda \theta^{-1} e^{-x/\theta}\, dx = \theta^\lambda \Gamma(1+\lambda).$$

Therefore, the condition of Proposition 4.8.1 is simply

$$\theta \leqslant \Gamma\big(2(\alpha+m)+3)\big)^{-1/2(\alpha+m+1)}.$$

It is natural to consider the asymptotic scale $e_i(t) = t^{-\alpha-i}$, with i nonnegative and less than α. We will derive only two terms, again not because of the difficulty of getting more, but for the space of writing the coefficients. So we assume α larger than 1 and derive a two terms expansion for the solution of (4.8.5).

In the chosen scale, $p_{\overline{K}} = (1, -\alpha)^t$. As we have seen in section 3.5, the matrix representing the derivative is defined by $\mathcal{D}\epsilon_i = -(\alpha+i)\epsilon_{i+1}$, that is

$$\mathcal{D} = \begin{pmatrix} 0 & 0 \\ -\alpha & 0 \end{pmatrix}.$$

The matrix representing the multiplication is

$$\mathcal{M}_c = c^\alpha \begin{pmatrix} 1 & 0 \\ 0 & c \end{pmatrix}.$$

The first moment of the Pareto distribution is $\mu_{K,1} = 1/(\alpha-1)$ while that of the exponential distribution is $\mu_{H,1} = \theta$. So the moment equation is

$$\mu_{F,1} = \mu_{K,1} + \mu_{H,1}\mu_{F,1} = (\alpha-1)^{-1} + \theta\mu_{F,1}.$$

Therefore,
$$\mu_{F,1} = 1/(1-\theta)(\alpha-1).$$

We then evaluate the matrices involved in (4.8.6). So,
$$\mathcal{L}_K = \begin{pmatrix} 1 & 0 \\ \alpha/(\alpha-1) & 1 \end{pmatrix}$$

and
$$\int \mathcal{M}_x \, dH(x) = \int \mathrm{diag}(x^\alpha, x^{\alpha+1}) \theta^{-1} e^{-x/\theta} \, dx$$
$$= \theta^\alpha \Gamma(\alpha+1) \begin{pmatrix} 1 & 0 \\ 0 & \theta(\alpha+1) \end{pmatrix}.$$

After some elementary calculation,
$$\mathcal{L}_{H \stackrel{\mathrm{M}}{\star} F} = \begin{pmatrix} 1 & 0 \\ \theta\alpha/(1-\theta)(\alpha-1) & 1 \end{pmatrix}.$$

Applying formula (4.8.6), we conclude that an asymptotic expansion for $\overline{F}$ is $p_{\overline{F},0} e_0 + p_{\overline{F},1} e_1$ with
$$p_{\overline{F}} = \begin{pmatrix} \dfrac{1}{1-\theta^\alpha \Gamma(\alpha+1)} \\ \alpha \dfrac{\theta^\alpha \Gamma(\alpha+1)(\theta-\alpha+\theta\alpha)+\alpha-1-\theta\alpha}{(1-\alpha)(1-\theta)(1-\theta^\alpha \Gamma(\alpha+1))(1-\theta^{\alpha+1}\Gamma(\alpha+2))} \end{pmatrix}.$$

To conclude this section, we mention that there are other parts of mathematics where regular variation plays an important role, e.g. differential equations; see for instance Marić and Tomić (1977), Omey (1981) and Marić (2000). Since the asymptotically smooth functions of fixed order can be differentiated a certain number of times, they form a natural class to use in differential equations. The general philosophy of this paper could be applied to some of these equations. Moreover, because of Theorem 2.5.1 it is possible to obtain asymptotic expansions for some integro-differential equations involving convolutions, Mellin transforms, multiplication by a function, nonlinearity due to taking powers of the unknown function and similar features for which we can find a stable asymptotic scale.

CHAPTER 5

Preparing the proof.

The purpose of this section is to provide more information on Laplace characters and smoothly varying functions of fixed order. Some of this material will be needed to prove Theorem 2.5.1.

5.1. Properties of Laplace characters.

We have seen that Laplace characters are invertible and we used their inverses. To write an explicit formula for the inverse, recall that an ordered partition of length k of an integer n is a k-tuple of positive integers $p = (p_1, \ldots, p_k)$ such that $p_1 \geqslant p_2 \geqslant \cdots \geqslant p_k > 0$ and $p_1 + \cdots + p_k = n$. The p_i are referred to as the parts of the partition. For such partition of length k, it is convenient for what follows to agree on the notation $p_i = 0$ for all i larger than k. We write $\mathcal{P}(n)$ for the set of all ordered partitions of n. For such partition p, we write Δp for $(p_1 - p_2, p_2 - p_3, \ldots, p_m - p_{m+1})$. Note that $(\Delta p)_i$ is the number of parts equal to i in the partition conjugate to p (see Stanley, 1999, §1.3 and §7.2). We also write $l(p)$ for the length of the partition, that is the number of p_i which are positive. Finally, $m(p) = (m_k(p))_{k \geqslant 0}$ denotes the counting function of the partition, whose k-th component, $m_k(p)$, counts the number of p_i equal to k. Also for a tuple $k = (k_1, \ldots, k_m)$ and an integer q we write $\binom{q}{k} = \binom{q}{k_1 \ldots k_m}$ the multinomial coefficient $q!/(k_1! \cdots k_m!)$.

The generating function for the cardinality of $\mathcal{P}(n)$ is $\prod_{i \geqslant 1}(1 - x^i)^{-1}$ (see, e.g., Stanley, 1999, §1.4).

PROPOSITION 5.1.1. *In $\mathbb{R}_m[\mathrm{D}]$, we have the inversion formula*

$$\mathcal{L}_{F,m}^{-1} = \sum_{0 \leqslant n \leqslant m} \Big(\sum_{p \in \mathcal{P}(n)} (-1)^{n+p_1} \binom{p_1}{\Delta p} \prod_{1 \leqslant k \leqslant m} \Big(\frac{\mu_{F,k}}{k!} \Big)^{(\Delta p)_k} \Big) \mathrm{D}^n$$

$$= \sum_{0 \leqslant n \leqslant m} \sum_{q \in \mathcal{P}(n)} (-1)^{n+l(q)} \binom{l(q)}{m(q)} \prod_{1 \leqslant k \leqslant m} \Big(\frac{\mu_{F,k}}{k!} \Big)^{m_k(q)} \mathrm{D}^n.$$

PROOF. In this proof we write $i = (i_1, \ldots, i_m)$ for a generic tuple of nonnegative integers, and $|i| = i_1 + \cdots + i_m$. In the space of formal power series in t,

$$\Big(1 + \sum_{1 \leqslant k \leqslant m} \frac{(-1)^k}{k!} \mu_{F,k} t^k\Big)^{-1} = \sum_{j \geqslant 0} (-1)^j \Big(\sum_{1 \leqslant k \leqslant m} \frac{(-1)^k}{k!} \mu_{F,k} t^k \Big)^j$$

$$= \sum_{j \geqslant 0} (-1)^j \sum_{|i|=j} \binom{j}{i} \prod_{1 \leqslant k \leqslant m} \Big(\frac{(-1)^k}{k!} \mu_{F,k} t^k \Big)^{i_k}.$$

65

Consequently, in $\mathbb{R}_m[\mathrm{D}]$, the inverse of $\mathcal{L}_{F,m}$ is

$$\sum_{0\leqslant j\leqslant m} (-1)^j \sum_i \binom{j}{i} \Bigl(\prod_{1\leqslant k\leqslant m} \Bigl(\frac{(-1)^k}{k!}\mu_{F,k}\Bigr)^{i_k}\Bigr) \mathrm{D}^{i_1+2i_2+\cdots+mi_m},$$

where the sum over i is over all tuples $(i_1,\ldots,i_m)$ such that $|i| = j$ and $i_1 + 2i_2 + \cdots + mi_m \leqslant m$. Set $p_{m+1} = 0$ and $p_l = i_l + p_{l+1}$ for $l = m, m-1, \ldots, 1$. Thus, $p_m = i_m$, $p_{m-1} = i_m + i_{m-1}$, $\ldots$, $p_1 = i_m + \cdots + i_1$. Set $p = (p_1,\ldots,p_m)$. We see that $\sum_{1\leqslant k\leqslant m} k i_k = n$ if and only if p is an ordered partition of n, and the correspondence between p and i is one-to-one. Moreover, $|i| = p_1$. Therefore, writing $\mathcal{P}(m,n)$ for the set of all ordered partitions of length at most m of n,

$$\mathcal{L}_{F,m}^{-1} = \sum_{n\leqslant m}\Bigl(\sum_{p\in\mathcal{P}(m,n)} (-1)^{p_1}\binom{p_1}{\Delta p}\prod_{1\leqslant k\leqslant m}\Bigl(\frac{(-1)^k}{k!}\mu_{F,k}\Bigr)^{(\Delta p)_k}\Bigr)\mathrm{D}^n.$$

Since $\mathcal{P}(m,n) = \mathcal{P}(n)$ whenever $n \leqslant m$, this is the first inversion formula.

The second formula is simply the first one where a partition p is replaced by its conjugate partition q. ∎

To use Proposition 5.1.1, one may need to generate all the partitions of a given integer. We refer to Nijenhuis and Wilf (1978) or Stanton and White (1986) for algorithms related to that matter.

Our next result is an alternative inversion formula for the Laplace characters. It is not as explicit as the previous one, because it is written in the ring $\mathbb{R}_m[\mathrm{D}]$. However, we used its matrix form in subsection 3.5.

PROPOSITION 5.1.2. *In $\mathbb{R}_m[\mathrm{D}]$, we have the inversion formula*

$$\mathcal{L}_{F,m}^{-1} = \sum_{0\leqslant k\leqslant m} (\mathrm{Id} - \mathcal{L}_{F,m})^k = \sum_{0\leqslant j\leqslant m}(-1)^j \mathcal{L}_{F,m}^j \sum_{j\leqslant k\leqslant m}\binom{k}{j}.$$

PROOF. Since $\mathcal{L}_{F,m} - \mathrm{Id}$ is in the ideal generated by D, it is nilpotent of nullity at most $m+1$ in $\mathbb{R}_m[\mathrm{D}]$. Consequently,

$$\mathcal{L}_{F,m}^{-1} = \bigl(\mathrm{Id} - (\mathrm{Id} - \mathcal{L}_{F,m})\bigr)^{-1} = \sum_{0\leqslant k\leqslant m}(\mathrm{Id} - \mathcal{L}_{F,m})^k.$$

Then, the second equality follows from the binomial formula

$$(\mathrm{Id} - \mathcal{L}_{F,m})^k = \sum_{0\leqslant j\leqslant k}\binom{k}{j}(-1)^j \mathcal{L}_{F,m}^j.$$
∎

We conclude this subsection with two lemmas. The first one is related to Proposition 2.3.1 while the second one describes the behavior of the Laplace characters under some multiplications. These lemmas are needed to prove Theorem 2.5.1.

5.2. PROPERTIES OF SMOOTHLY VARYING FUNCTIONS OF FINITE ORDER

LEMMA 5.1.3. *If H and K are two distribution functions with finite m-th absolute moments, then*

$$L_{K \star H, m} = \sum_{0 \leqslant j \leqslant m} \frac{(-1)^j}{j!} \mu_{H,j} L_{K, m-j} \mathrm{D}^j.$$

PROOF. The right hand side of the equality is

$$\sum_{0 \leqslant j \leqslant m} \frac{(-1)^j}{j!} \mu_{H,j} \sum_{0 \leqslant l \leqslant m-j} \frac{(-1)^l}{l!} \mu_{K,l} \mathrm{D}^{j+l}.$$

Setting $s = l + j$, it is

$$\sum_{s,j \geqslant 0} \mathbb{1}\{j \leqslant s \leqslant m\} \frac{(-1)^s}{s!} \binom{s}{j} \mu_{H,j} \mu_{K,s-j} \mathrm{D}^s = \sum_{0 \leqslant s \leqslant m} \frac{(-1)^s}{s!} \mu_{K \star H, s} \mathrm{D}^s,$$

which gives the conclusion. ∎

LEMMA 5.1.4. *If K is a distribution function whose m-th moment is finite, then*

$$L_{M_\lambda K, m} M_\lambda = M_\lambda L_{K, m}.$$

PROOF. The result follows from the equalities

$$L_{M_\lambda K, m}(M_\lambda h) = \sum_{0 \leqslant j \leqslant m} \frac{(-1)^j}{j!} \mu_{M_\lambda K, j} \mathrm{D}^j (M_\lambda h)$$

$$= \sum_{0 \leqslant j \leqslant m} \frac{(-1)^j}{j!} \lambda^j \mu_{K,j} \lambda^{-j} M_\lambda \mathrm{D}^j h,$$

and the equality of the last term with $M_\lambda L_{K,m} h$. ∎

5.2. Properties of smoothly varying functions of finite order.

In order to check that a function is smoothly varying of a given order ω, we mentioned in section 2.4 that it suffices to show that its $\lfloor \omega \rfloor + 1$-th derivative is regularly varying. The first result of this subsection proves that assertion, showing that the spaces $SR_{-\alpha, \omega}$, $\omega \geqslant 0$, are nested.

PROPOSITION 5.2.1. *If r is less than s, then $SR_{-\alpha, s} \subset SR_{-\alpha, r}$.*

PROOF. If r or s is an integer, the result is obvious; hence we assume that both are not integers. Write $r = m + \rho$ and $s = n + \sigma$ with m and n integers and ρ, σ positive and less than 1. Let h be a function in $SR_{-\alpha, s}$. If n is at most r, then m

and n are equal, and ρ is less than σ. It is then clear that h is in $SR_{-\alpha,r}$, since the function $|x|^{\sigma-\rho}$ is bounded in any neighborhood of the origin.

Assume that n is larger than r. Then n is at least $m+1$ and

$$\lim_{t\to\infty} t^{m+1} h^{(m+1)}(t)/h(t) = (-\alpha)_{m+1}.$$

Since $m+1$ is at least 1, the function $h^{(m+1)}$ is regularly varying of index $-\alpha-m-1$. We write

$$h^{(m)}\big(t(1-x)\big) - h^{(m)}(t) = -\int_{1-x}^{1} t h^{(m+1)}(tu)\,du$$

to obtain

$$|\Delta_{t,x}^{\rho} h^{(m)}| \leqslant |x|^{1-\rho} \sup_{1-x \leqslant u \leqslant 1} |t h^{(m+1)}(tu)/h^{(m)}(t)|.$$

Using the uniform convergence theorem (Bingham, Goldie and Teugels, 1989, Theorem 1.2.1) and that ρ is less than 1, we conclude that

$$\lim_{\delta\to 0} \limsup_{t\to\infty} \sup_{0<|x|\leqslant\delta} |\Delta_{t,x}^{\rho} h^{(m)}| = 0,$$

which shows that h belongs to $SR_{-\alpha,r}$. ∎

In view of Proposition 5.2.1, one may wonder if $SR_{-\alpha,s}$ is equal to $\cup_{r>s} SR_{-\alpha,r}$. It is not in general, and the following example proves that to be the case when s is an integer. Consider the function g defined on the nonnegative half-line by

$$g(x) = |x-n|^{1/n} \quad \text{on } [n-1/4, n+1/4],\ n \geqslant 1$$

and interpolated linearly outside those intervals. Take $h(t) = t^{-\alpha} + t^{-\alpha-1} g(t)$. Because g is bounded, h is regularly varying with index $-\alpha$ and belongs to $SR_{-\alpha,0}$. But

$$|\Delta_{t,x}^{r} h| = \frac{1}{1+t^{-1}g(t)} \frac{(1-x)^{-\alpha}-1}{|x|^r} \tag{5.2.1}$$

$$+ \frac{1}{t+g(t)} (1-x)^{-\alpha-1} \frac{g\big(t(1-x)\big)-g(t)}{|x|^r} \tag{5.2.2}$$

$$+ \frac{1}{t+g(t)} \frac{(1-x)^{-\alpha-1}-1}{|x|^r} g(t). \tag{5.2.3}$$

To see if h belongs to $SR_{-\alpha,r}$, one needs to take the supremum of $|\Delta_{t,x}^{r} h|$ for $|x|$ positive at most δ, and then the supremum limit as t tends to infinity. Taking these two limits in terms (5.2.1) and (5.2.3) yields 0. But, when t is an integer n and x is, say, y/n with $|y|$ at most $1/4$, the last fraction in (5.2.2) is

$$\frac{n^r g(n-y)}{|y|^r} = n^r y^{1/n-r}.$$

The supremum over y is infinite as soon as $n > 1/r$. Hence, the function h does not belong to any $SR_{-\alpha,\epsilon}$ for some positive ϵ. Integrating m times the function h gives a function in $SR_{-\alpha,m}$ which does not belong to $\cup_{r>m} SR_{-\alpha,r}$.

Among the regularly varying functions, the normalized ones will be of importance to us. Following Bingham et al. (1989, §1.3.2), we say that a function g defined on $[a, \infty)$ is a normalized regularly varying function with index ρ if it has the representation

$$g(t) = t^\rho c \exp \int_a^t \frac{\epsilon(u)}{u} \, du, \qquad (5.2.4)$$

where $\epsilon(\cdot)$ is a function converging to 0 at infinity.

If g is a function such that $\mathrm{Id}\, g'/g$ has a negative limit $-\alpha$ at infinity, then g is regularly varying with index $-\alpha$. Therefore, if h belongs to $SR_{-\alpha,m}$, then all $h^{(k)}$'s for $k = 0, 1, \ldots, m-1$ are normalized regularly varying. Consequently, for any such k, the function $t \mapsto t^\sigma |h^{(k)}(t)|$ is ultimately decreasing (respectively increasing) when $\sigma < \alpha + k$ (respectively $\sigma > \alpha + k$) — see Bingham, Goldie and Teugels, 1989, Theorem 1.5.5.

Before going further, let us make a digression about condition (2.4.2). One may wonder about the analogous condition when r is 1. In fact, as we will explicate next, the condition (2.4.2) with $r = 1$ has bearing on several issues, e.g. understanding the spaces $SR_{-\alpha,\omega}$, the monotone density theorem which is a classical result in the theory of regular variation, the class of asymptotically smooth functions introduced in Barbe and McCormick (2005) as well as the conditions [D] in Borovkov and Borovkov (2003).

For clarity of the argument, if a function g is differentiable, then

$$\lim_{x \to 0} \Delta^1_{t,x} g = -t g'(t)/g(t).$$

Therefore, if β is a positive number and if g' is regularly varying with index $-\beta - 1$,

$$\lim_{t \to \infty} \lim_{x \to 0} \Delta^1_{t,x} g = \beta.$$

For $r = 1$, it is then natural to introduce the condition

$$\lim_{\delta \to 0} \limsup_{t \to \infty} \sup_{0 < |x| < \delta} |\Delta^1_{t,x} g - \beta| = 0. \qquad (5.2.5)$$

PROPOSITION 5.2.2. *The following are equivalent:*
(i) condition (5.2.5).
(ii) g is normalized regularly varying with index $-\beta$.
(iii) g is ultimately absolutely continuous and a version of its Radon-Nikodým derivative is regularly varying with index $-\beta - 1$.

PROOF. We write $\dot{g}$ for a Radon-Nikodým derivative of g when it exists.

We start by proving that (iii) implies (ii). If (iii) holds, by Karamata's theorem (see Bingham et al., 1989, Theorem 1.5.11), $\mathrm{Id}\, \dot{g}/g \sim -\beta$ at infinity. Integrating $\dot{g}/g$ gives (ii).

Next, we show that (ii) implies (iii). Consider the Karamata representation of g,

$$g(t) = t^{-\beta} c \exp \int_{t_0}^{t} \frac{\epsilon(u)}{u}\, du\,.$$

Such a function is ultimately absolutely continuous and a Radon-Nikodým derivative is

$$\dot{g}(t) = \frac{g(t)}{t}\left(-\beta + \frac{\epsilon(t)}{\beta}\right).$$

This implies (iii).

We now prove that (ii) implies (i). Again, the Karamata representation of g yields for t large enough and x small enough,

$$\Delta^1_{t,x} g = \frac{(1-x)^{-\beta} - 1}{x} + (1-x)^{-\beta} \frac{1}{x} \left(\exp \int_{t}^{t(1-x)} \frac{\epsilon(u)}{u}\, du - 1 \right).$$

Let η be a positive number and let t_1 be large enough so that g is absolutely continuous on (t_1, ∞) and the absolute value of $\epsilon(\cdot)$ is at most η on (t_1, ∞). Then, for any t and $t(1-x)$ more than t_1,

$$-\eta |\log(1-x)| \leqslant \int_{t}^{t(1-x)} \frac{\epsilon(u)}{u}\, du \leqslant \eta |\log(1-x)|\,.$$

For δ small enough, this implies

$$\limsup_{t\to\infty} \sup_{0<|x|<\delta} \left| \frac{1}{x}\left(\exp \int_{t}^{t(1-x)} \frac{\epsilon(u)}{u}\, du - 1 \right) \right| \leqslant 2\eta\,.$$

On the other hand,

$$\lim_{\delta \to 0} \limsup_{t\to\infty} \sup_{0<|x|<\delta} \left| \frac{(1-x)^{-\beta} - 1}{x} - \beta \right| = 0\,.$$

Therefore, (5.2.5) and (i) hold.

We conclude the proof of the Proposition by showing that (i) implies (ii). We first note that if (5.2.5) holds, then g must be ultimately continuous; otherwise, the supremum in x in (5.2.5) would be infinite for some arbitrarily large t's, precluding (5.2.5) to hold. Also, g does not vanish ultimately. So, without loss of generality, we assume that g is ultimately positive. Let $\dot{g}_U$ and $\dot{g}_L$ be the upper and lower derivatives of g, that is

$$\dot{g}_U(t) = \limsup_{\epsilon \to 0} \frac{g(t+\epsilon) - g(t)}{\epsilon},$$

$$\dot{g}_L(t) = \liminf_{\epsilon \to 0} \frac{g(t+\epsilon) - g(t)}{\epsilon}.$$

Condition (5.2.5) implies that $\mathrm{Id}\,\dot{g}_U/g$ and $\mathrm{Id}\,\dot{g}_L/g$ have limit $-\beta$ at infinity. Set $f(t)=t^\beta g(t)$. Since g is ultimately continuous,

$$\dot{f}_U(t)=\beta t^{\beta-1}g(t)+t^\beta \dot{g}_U(t)\,,$$

and an analogous relation holds for the lower derivative of f. This implies

$$t\dot{f}_U(t)/f(t)=\beta+t\dot{g}_U(t)/g(t)$$

has limit 0 at infinity, and similarly for the upper derivative. By Bojanic and Karamata (1963) — see Bingham et al., 1989, exercise 1.11.8 — and Theorem 1.5.5 in Bingham et al. (1989), this implies that f is normalized slowly varying. ∎

Note that Proposition 5.2.2 implies that if a function h belongs to $SR_{-\alpha,m}$ and (5.2.5) holds with $h^{(m)}$ in place of g and $\beta=\alpha+m$, then the existence of the derivative $h^{(m+1)}$ guarantees that h belongs to $SR_{-\alpha,m+1}$. However, without assuming that $h^{(m+1)}$ exists, (5.2.5) would only give existence and regular variation of a Radon-Nikodým derivative of $h^{(m)}$.

The equivalence between (ii) and (iii) in Proposition 5.2.2 shows that the class of normalized regularly varying functions is a very natural one when dealing with regular variation of a function and its derivative.

It also follows from Proposition 5.2.2 that the class of asymptotically smooth distributions in Barbe and McCormick (2005) is in fact those which are normalized regularly varying; this was suggested to us by Jaap Geluk and prompted Proposition 5.2.2. It also shows that condition $[D_m]$ of Borovkov and Borovkov (2003) relates to normalized regular variation.

Proposition 5.2.2 is also interesting with respect to the monotone density theorem (Bingham et al., 1989, Theorem 1.7.2). Indeed, this theorem asserts that if a regularly varying function g with index $-\beta$ has ultimately monotone derivative g', then this derivative is regularly varying of index $-\beta-1$; and moreover, by Karamata's theorem, $\mathrm{Id}\,g'/g \sim -\beta$. This implies that g is normalized regularly varying, and therefore satisfies (ii) of Proposition 5.2.2. In other words, the monotone density theorem can be viewed as a particular case of the implication of (iii) by (ii) in Proposition 5.2.2. This concludes our digression on condition (2.4.2).

The importance of the classes $SR_{-\alpha,m}$ in this paper stems from the nice behavior of their functions with respect to differentiation, global Potter type bounds, and to Taylor's formula used asymptotically. In order to elaborate on this assertion, recall first that Potter's bounds (see, Bingham et al., 1989, Theorem 1.5.6) assert that if g is a function which is regularly varying with index ρ, then for any A larger than 1 and any δ positive, there exists a t_0 such that for any λ at least 1 and any t more than t_0,

$$A^{-1}\lambda^{\rho-\delta}\leqslant g(\lambda t)/g(t)\leqslant A\lambda^{\rho+\delta}\,.$$

This standard Potter inequality already yields an upper bound on the decay of the scaled derivatives of smoothly varying functions of fixed order.

LEMMA 5.2.3. *Let h be a smoothly varying function of index $-\alpha$ and order s more than 1. Let ϵ be a positive number. There exists t_1 such that for any c positive at most 1 and any t at least t_1 and any integer k at most s,*

$$|(M_c h)^{(k)}(t)| \leqslant 2|(-\alpha)_k| c^{\alpha-\epsilon} t^{-k} |h(t)|.$$

PROOF. We have $(M_c h)^{(k)}(t) = c^{-k} h^{(k)}(t/c)$. Since h is in $SR_{-\alpha,s}$ with s at least k, there exists t_0 such that for any t at least t_0,

$$|t^k h^{(k)}(t)/h(t)| \leqslant \sqrt{2} |(-\alpha)_k|.$$

Since c is positive and at most 1, the inequality $t \geqslant t_0$ implies $t/c \geqslant t_0$. Therefore, for t at least t_0,

$$c^{-k} |h^{(k)}(t/c)| \leqslant \sqrt{2} |(-\alpha)_k| t^{-k} |h(t/c)|.$$

Now, if t is larger than some t_1 independent of c, Potter's bounds give the inequality $|h(t/c)| \leqslant \sqrt{2} c^{\alpha-\epsilon} |h(t)|$. This implies the result. ∎

The following lemma shows that for normalized regularly varying functions, in particular for smoothly varying functions of fixed order at least 1, Potter's bounds can be improved by taking $A = 1$. Although this may seem a minor feature, this improvement is essential for our purposes.

LEMMA 5.2.4. *Let g be a normalized regularly varying function with index ρ. Let δ be a positive number. There exists t_2 such that for any λ larger than 1 and any t at least t_2,*

$$\lambda^{\rho-\delta} \leqslant g(t\lambda)/g(t) \leqslant \lambda^{\rho+\delta}.$$

PROOF. Let $\epsilon(\cdot)$ be as in the Karamata representation (5.2.4). Let t_2 be at least a and such that $\sup_{t \geqslant t_2} |\epsilon(t)| \leqslant \delta$. Then,

$$\frac{g(\lambda t)}{g(t)} = \lambda^\rho \exp\left(\int_t^{t\lambda} \frac{\epsilon(u)}{u} du\right) \begin{cases} \leqslant \lambda^\rho \exp(\delta \int_t^{t\lambda} u^{-1} du) \\ \geqslant \lambda^\rho \exp(-\delta \int_t^{t\lambda} u^{-1} du), \end{cases}$$

which yields the result. ∎

To connect Taylor's formula with asymptotic expansions, we set

$$\overline{\Delta}_{\tau,\delta}^q(h) = \sup_{t \geqslant \tau} \sup_{0 < |x| \leqslant \delta} |\Delta_{t,x}^q h|.$$

PROPOSITION 5.2.5. *Let r be in $[0,1]$. If h is m times differentiable, then for any positive t and any u,*

$$\left| h(t+u) - \sum_{0 \leqslant j \leqslant m} \frac{u^j}{j!} h^{(j)}(t) \right| \leqslant \frac{|u|^{m+r}}{t^r} \frac{|h^{(m)}(t)|}{m!} \overline{\Delta}_{t,|u|/t}^r (h^{(m)}).$$

5.2. PROPERTIES OF SMOOTHLY VARYING FUNCTIONS OF FINITE ORDER

PROOF. The Taylor-McLaurin formula yields

$$h(t+u) = \sum_{0 \leqslant j \leqslant m} \frac{u^j}{j!} h^{(j)}(t) + \frac{u^m}{m!} \left(h^{(m)}(t + \theta_{t,u} u) - h^{(m)}(t) \right)$$

with $\theta_{t,u}$ between 0 and 1. The equality

$$|h^{(m)}(t + \theta_{t,u} u) - h^{(m)}(t)| = \theta_{t,u}^r t^{-r} |u|^r |h^{(m)}(t)| |\Delta_{t,-\theta_{t,u} u/t}^r h^{(m)}|$$

implies the result. ∎

Observe that Lemma 5.2.3 asserts that if h is smoothly varying of order $s = m+r$, then $h^{(m)}(t)$ is of order $t^{-m} h(t)$. Thus, the remainder term in Proposition 5.2.5 is expected to be asymptotically rather small and certainly of smaller order of magnitude than any of the $h^{(j)}(t)$ involved in the inequality of Proposition 5.2.5. In other words, for functions in $SR_{-\alpha,s}$ we can relate the local character of Taylor's formula to an asymptotic expansion of the translation $u \mapsto t + u$ acting on those functions. Proposition 5.2.5 implies that if h is in $SR_{-\alpha,s}$, with s larger than m, then $\sum_{0 \leqslant j \leqslant m} (u^j/j!) h^{(j)}(t)$ is an asymptotic expansion of $h(t+u)$ as t tends to infinity. Using the translation τ_x defined on functions by $\tau_x h(t) = h(t+x)$, Proposition 5.2.5 asserts that τ_u is approximately the Laplace character $L_{\delta_u, m}$ when read on the tail of smoothly varying functions.

Before moving to the next section, we make a remark concerning the behavior of the operator $\Delta_{t,x}^r$ with respect to composition by differentiation and multiplications M_c. This remark does not have much intrinsic value, but we will need it during our proof.

LEMMA 5.2.6. *For any positive c, the equality $\Delta_{t,x}^r D^j M_c = \Delta_{t/c,x}^r D^j$ holds.*

PROOF. We have

$$D^j M_c h(t) = \frac{d^j h(t/c)}{dt^j} = c^{-j} h^{(j)}(t/c).$$

Therefore,

$$\Delta_{t,x}^r D^j M_c h = \text{sign}(x) \frac{h^{(j)}\big((t/c)(1-x)\big) - h^{(j)}(t/c)}{|x|^r h^{(j)}(t/c)} = \Delta_{t/c,x}^r D^j h. \qquad \blacksquare$$

CHAPTER 6

Proof in the positive case.

In this section, we prove Theorem 2.5.1 under some extra assumptions. Throughout this section, we suppose that $F^{(k)}$ is bounded and Lebesgue integrable over the positive half-line. More importantly, we also suppose that both the c_i's and the X_i's are nonnegative. Since $\mathbb{Z}$ and $\mathbb{N}^*$ are in bijection, there is no loss of generality to index our sequence (c_i) by $\mathbb{N}^*$, which we will do throughout the proof. In the first subsection, we derive an expression for convolutions which will be suitable for our analysis. In the second subsection, we outline the proof. The third subsection contains some basic facts about regularly varying functions and tail estimates. Subsection 4 contains a key estimate. Then we prove some basic lemmas (subsection 5) and finally conclude the proof by induction.

There will be many results during the proof which state that some term A say is at most some term B. In some cases, nothing prevents a priori B to be infinite. However, at the end, we will see that all the upper bounds are indeed finite and prove the theorem.

During the proof, M denotes a generic constant which may change from place to place. This constant 'depends' only on F and ω. Also several of our lemmas below have conclusions which hold for all sufficiently large reals. The bounds above which these conclusions hold are denoted by t_i. These t_i's only depend on F and ω.

Since we consider nonnegative c_i's throughout this section, we assume without any loss of generality that the sequence $(c_n)_{n \geqslant 1}$ is nonincreasing. It is also understood that except in subsection 6.1, all distributions in the current section are supported by the nonnegative half-line. Recall the notation C_s for the sum of the c_i^s, that is, in this section, $C_s = \sum_{i \geqslant 1} c_i^s$.

Note also that since ω is at least 1, the function $\overline{F}$ is normalized regularly varying.

6.1. Decomposition of the convolution into integral and multiplication operators.

Let K be a distribution function and h be a function integrable with respect to the measure dK. For any η between 0 and 1, define the operator

$$T_{K,\eta}h(t) = \int_{-\infty}^{\eta t} h(t-x)\,\mathrm{d}K(x).$$

Recall that for any positive real number c, we defined the multiplication operator on functions

$$M_c h(t) = h(t/c).$$

75

It is then natural to define, when possible,

$$M_0 h(t) = \begin{cases} \lim_{s \to +\infty} h(s) & \text{if } t \geqslant 0, \\ \lim_{s \to -\infty} h(s) & \text{if } t < 0. \end{cases}$$

Note that if a random variable X has distribution function F, then cX has distribution function $M_c F$. Moreover, for any c nonnegative, $\overline{M_c F} = M_c \overline{F}$.

These operators allow us to write convolutions and their derivatives in a way that will be suitable for our analysis.

PROPOSITION 6.1.1. *Let F and G be two distribution functions. For any η between 0 and 1,*

$$\overline{F \star G} = T_{G, 1-\eta} \overline{F} + T_{F, \eta} \overline{G} + M_{1/\eta} \overline{F} M_{1/(1-\eta)} \overline{G}.$$

Let k be a positive integer. If F and G are k times differentiable,

$$\overline{F \star G}^{(k)} = T_{G, 1-\eta} \overline{F}^{(k)} + T_{F, \eta} \overline{G}^{(k)} - \sum_{1 \leqslant i \leqslant k-1} M_{1/\eta} \overline{F}^{(i)} M_{1/(1-\eta)} \overline{G}^{(k-i)}.$$

When $k = 1$, the sum $\sum_{1 \leqslant i \leqslant 0}$ in the proposition must be read as 0. Equivalently,

$$(F \star G)' = T_{G, 1-\eta} F' + T_{F, \eta} G'.$$

The proof of the proposition will be based on a lemma describing the behavior of the operators $T_{K, \eta}$ and M_c under differentiation.

LEMMA 6.1.2. *If K is a differentiable distribution function and if h is a differentiable function, then*

$$(T_{K, \eta} h)' = T_{K, \eta} h' + \eta M_{1/(1-\eta)} h M_{1/\eta} K'$$

and

$$(M_c h)' = c^{-1} M_c h'.$$

PROOF. If K is differentiable, then

$$T_{K, \eta} h(t) = \int_{-\infty}^{\eta t} h(t - x) K'(x) \, dx.$$

The chain rule yields

$$\frac{d}{dt} T_{K, \eta} h(t) = \int_{-\infty}^{\eta t} h'(t - x) K'(x) \, dx + \eta h\big(t(1 - \eta)\big) K'(\eta t)$$

which is the first statement. The second one is immediate. ∎

PROOF (of the Proposition). The tail of the convolution $F * G$ is

$$\overline{F \star G}(t) = \int_{-\infty}^{\infty} \overline{F}(t-y) \, dG(y) \qquad (6.1.1)$$
$$= \int_{-\infty}^{t(1-\eta)} \overline{F}(t-y) \, dG(y) + \int_{t(1-\eta)}^{\infty} \overline{F}(t-y) \, dG(y).$$

We integrate by parts and make a change of variable to obtain

$$\int_{t(1-\eta)}^{\infty} \overline{F}(t-y) \, dG(y) = \left[\overline{F}(t-y)(G(y)-1) \right]_{t(1-\eta)}^{\infty} + \int_{t(1-\eta)}^{\infty} \overline{G}(y) \, d\overline{F}(t-y)$$
$$= \overline{F}(t\eta)\overline{G}(t(1-\eta)) + \int_{-\infty}^{t\eta} \overline{G}(t-y) \, dF(y).$$

Combined with (6.1.1), we obtain the first assertion of the proposition.

To prove the second assertion, we proceed by induction, starting to prove the result for $k=1$. Using the lemma and differentiating the expression for $\overline{F \star G}$, we see that
$$\overline{F \star G}' = T_{G,1-\eta}\overline{F}' + T_{F,\eta}\overline{G}' + (1-\eta)M_{1/\eta}\overline{F}M_{1/(1-\eta)}G'$$
$$+ \eta M_{1/(1-\eta)}\overline{G}M_{1/\eta}F' + \eta M_{1/\eta}\overline{F}'M_{1/(1-\eta)}\overline{G}$$
$$+ (1-\eta)M_{1/\eta}\overline{F}M_{1/(1-\eta)}\overline{G}'.$$

Since $\overline{F}' = -F'$, we obtain $\overline{F \star G}' = T_{G,1-\eta}\overline{F}' + T_{F,\eta}\overline{G}'$.

Assume now that the relation holds for any integer up to $k-1$. Using the lemma,

$$(\overline{F \star G}^{(k-1)})' = T_{G,1-\eta}\overline{F}^{(k)} + T_{F,\eta}\overline{G}^{(k)}$$
$$+ (1-\eta)M_{1/\eta}\overline{F}^{(k-1)}M_{1/(1-\eta)}G'$$
$$+ \eta M_{1/(1-\eta)}\overline{G}^{(k-1)}M_{1/\eta}F'$$
$$- \sum_{1\leqslant i \leqslant k-2} \eta M_{1/\eta}\overline{F}^{(i+1)}M_{1/(1-\eta)}\overline{G}^{(k-1-i)}$$
$$- \sum_{1\leqslant i \leqslant k-2} (1-\eta)M_{1/\eta}\overline{F}^{(i)}M_{1/(1-\eta)}\overline{G}^{(k-i)},$$

which after collecting the terms gives the proper expression for $\overline{F \star G}^{(k)}$. ∎

6.2. Organizing the proof.

The proof goes essentially by induction on k, and for fixed k by induction on n. We write G_n for the distribution function of $\sum_{1\leqslant i \leqslant n} c_i X_i$ and G for that of $\langle c, X \rangle$. We also define $G_{n\setminus i}$ and $G_{\setminus i}$ to be respectively the distribution functions of

$$\sum_{\substack{1\leqslant j \leqslant n \\ j \neq i}} c_j X_j \qquad \text{and} \qquad \sum_{\substack{j \geqslant 1 \\ j \neq i}} c_j X_j.$$

We also write F_i for $M_{c_i}F$, that is for the distribution function of c_iX_i. Hence, $G_n = F_1 \star \cdots \star F_n$.

We also let ρ denote a parameter in $(0,1)$, whose value will be set at the end of the proof. We set $d_n = c_n^\rho$.

Applying Proposition 6.1.1, we see that

$$\overline{G}_n^{(k)} = T_{G_{n-1},1-d_n}\overline{F}_n^{(k)} + T_{F_n,d_n}\overline{G}_{n-1}^{(k)} - \sum_{1\leqslant i \leqslant k-1} M_{1/d_n}\overline{F}_n^{(i)}M_{1/(1-d_n)}\overline{G}_{n-1}^{(k-i)}.$$

Let us write $\mathcal{A}_m\overline{G}_n^{(k)}$ for the m terms asymptotic expansion of $\overline{G}_n^{(k)}$ — or, more precisely, for the time being, the candidate for this asymptotic expansion —

$$\mathcal{A}_m\overline{G}_n^{(k)} = \sum_{1\leqslant i \leqslant n} L_{G_{n\setminus i},m}\overline{F}_i^{(k)}.$$

The '$\mathcal{A}_m$' in this formula is not an operator; it is a short hand for 'm-terms approximation of' and merely to help the memory. Sometimes we will omit the subscript m, writing simply $\mathcal{A}$. Similarly, we will often drop the subscript m indicating the order of a Laplace character when there is no ambiguity.

We will approximate $T_{G_{n-1},1-d_n}$ by $L_{G_{n-1}}$ (Theorem 6.4.1) and T_{F_n,d_n} by L_{F_n}. Thus, we expect an approximation of $\overline{G}_n^{(k)}$ by $L_{G_{n-1}}\overline{F}_n^{(k)} + L_{F_n}\overline{G}_{n-1}^{(k)}$. Since $\overline{G}_{n-1}^{(k)}$ is close to its asymptotic expansion, we also hope to approximate $L_{F_n}\overline{G}_{n-1}^{(k)}$ by $L_{F_n}\mathcal{A}\overline{G}_{n-1}^{(k)}$. But

$$L_{F_n}\mathcal{A}\overline{G}_{n-1}^{(k)} = \sum_{1\leqslant i \leqslant n-1} L_{F_n}L_{G_{n-1\setminus i}}\overline{F}_i^{(k)}.$$

Now, $L_{F_n}L_{G_{n-1\setminus i}}$ is not $L_{F_n\star(G_{n-1\setminus i})}$ (the latter being $L_{G_{n\setminus i}}$); however, these two operators are equal up to differential operators involving D^{m+1+i}'s for i nonnegative. Since $\overline{F}_i^{(k+l)} \asymp \mathrm{Id}^{-l}\overline{F}_i^{(k)}$ when $\overline{F}_i^{(k+l)}$ is regularly varying, we expect that

$$L_{F_n}\mathcal{A}\overline{G}_{n-1}^{(k)} \sim \sum_{1\leqslant i \leqslant n-1} L_{G_{n\setminus i}}\overline{F}_i^{(k)}.$$

Hence, we should obtain

$$\overline{G}_n^{(k)} \sim L_{G_{n-1}}\overline{F}_n^{(k)} + \sum_{1\leqslant i \leqslant n-1} L_{G_{n\setminus i}}\overline{F}_i^{(k)},$$

which is equal to $\mathcal{A}\overline{G}_n^{(k)}$. Thus, assuming that $\overline{G}_{n-1}^{(k)} \sim \mathcal{A}\overline{G}_{n-1}^{(k)}$, we can expect to prove $\overline{G}_n^{(k)} \sim \mathcal{A}\overline{G}_n^{(k)}$, and build in this way a proof by induction on n. In any case, this derivation shows that, for k at least 2,

$$\begin{aligned}\overline{G}_n^{(k)} - \mathcal{A}\overline{G}_n^{(k)} &= (T_{G_{n-1},1-d_n} - L_{G_{n-1}})\overline{F}_n^{(k)} \\ &+ T_{F_n,d_n}(\overline{G}_{n-1}^{(k)} - \mathcal{A}\overline{G}_{n-1}^{(k)}) \\ &+ \left(T_{F_n,d_n}\mathcal{A}\overline{G}_{n-1}^{(k)} - \sum_{1\leqslant i \leqslant n-1} L_{G_{n\setminus i}}\overline{F}_i^{(k)}\right) \\ &- \sum_{1\leqslant j \leqslant k-1} M_{1/d_n}\overline{F}_n^{(j)}M_{1/1-d_n}\overline{G}_{n-1}^{(k-j)}.\end{aligned} \quad (6.2.1)$$

A similar equality holds when k is 0 or 1.

In the right hand side of this equality, we control $\overline{F}_n^{(k)}$ by assumptions on $\overline{F}$ and the difference $\mathcal{T}_{G_{n-1}, 1-d_n} - \mathcal{L}_{G_{n-1}}$ by Theorem 6.4.1. Since $\mathcal{A}\overline{G}_{n-1}^{(k)}$ is expressed in terms of functions $\overline{F}_i$'s, and hence, ultimately in terms of $\overline{F}$, it is an explicit term on which we can work. The only term on which we do not have a rather direct control is $\mathcal{T}_{F_n, d_n}(\overline{G}_{n-1}^{(k)} - \mathcal{A}\overline{G}_{n-1}^{(k)})$. Up to a shift of index n, it is related to the left hand side of (6.2.1), provided we can control the operator norm of $\mathcal{T}_{F_n, d_n}$. This will allow us to prove the asymptotic expansion $\overline{G}_n^{(k)} \sim \mathcal{A}_m \overline{G}_n^{(k)}$ is valid uniformly in n.

For the induction and the estimates to be written in a manageable form, we use a pseudo-semi-norm to control the remainder terms. For any function h on the nonnegative half-line, we define

$$|h|_{m,\tau} = \sup_{t \geqslant \tau} t^m |h(t)|/\overline{F}(t).$$

The choice of this pseudo-semi-norm is motivated by the fact that $\lim_{\tau \to \infty} |h|_{m,\tau} = 0$ is equivalent to $h(t) = o(t^{-m}\overline{F}(t))$ as t tends to infinity. Therefore, proving an asymptotic equivalence up to a $t^{-m}\overline{F}(t)$ term amounts to proving that the pseudo-semi-norm $|\cdot|_{m,\tau}$ of some function tends to 0 as τ tends to infinity.

6.3. Regular variation and basic tail estimates.

In this subsection, we collect some facts on regular variation and on the distribution of $\langle c, X \rangle$. Our first lemma sandwiches the tail of $\langle c, X \rangle$.

LEMMA 6.3.1. *If $C_\rho \leqslant 1$, then pt*

$$M_{c_1}\overline{F} \leqslant \overline{G}_n \leqslant \overline{G} \leqslant \sum_{n \geqslant 1} M_{c_n^{1-\rho}}\overline{F}.$$

Moreover, let ϵ be a positive real number. There exists t_3 such that for any t larger than t_3,

$$\overline{G}(t) \leqslant C_{\alpha(1-\rho)-\epsilon}\overline{F}(t).$$

PROOF. The first two bounds follow from the inequalities

$$c_1 X_1 \leqslant c_1 X_1 + \cdots + c_n X_n \leqslant \langle c, X \rangle.$$

Next, if $c_n^{1-\rho} X_n \leqslant t$ for all n, then pt

$$\sum_{n \geqslant 1} c_n X_n \leqslant t \sum_{n \geqslant 1} c_n^\rho = tC_\rho \leqslant t.$$

Therefore,

$$P\{\langle c, X \rangle > t\} \leqslant P\{\cup_{n \geqslant 1}\{c_n^{1-\rho} X_n > t\}\} = \sum_{n \geqslant 1} P\{c_n^{1-\rho} X_n > t\},$$

which is the last inequality in the first statement. This inequality asserts that $\overline{G}(t)$ is at most $\sum_{n\geqslant 1} \overline{F}(t/c_n^{1-\rho})$. Since C_ρ is at most 1, so are all the c_i's. Apply Lemma 5.2.4 with $g = \overline{F}$, $\delta = \epsilon/(1-\rho)$ and $\lambda = c_n^{\rho-1}$ to obtain the second statement. ∎

This lemma will be useful when combined with the next analytical result.

LEMMA 6.3.2. *Let g, h be two nonincreasing functions of bounded variations, such that $g \leqslant h$ and*
$$\lim_{t\to\infty} g(t) = \lim_{t\to\infty} h(t) = 0.$$
Let f be a nonnegative nondecreasing and right continuous function on $\mathbb{R}^+$. Then
$$-\int f\,dg \leqslant -\int f\,dh.$$

PROOF. Define the left continuous inverse of f by
$$f^{\leftarrow}(u) = \inf\{\,x \,:\, f(x) \geqslant u\,\}.$$
Observe that $f(x) \geqslant u$ if and only if $f^{\leftarrow}(u) \leqslant x$. We use the Fubini-Tonelli theorem to write
$$\int_0^\infty f\,dg = \int_0^\infty \int_0^\infty \mathbb{1}\{\,u \leqslant f(x)\,\}\,du\,dg(x)$$
$$= \int_0^\infty \int_0^\infty \mathbb{1}\{\,x \geqslant f^{\leftarrow}(u)\,\}\,dg(x)\,du$$
$$= -\int_0^\infty g \circ f^{\leftarrow}(u)\,du.$$
The result follows from the inequality $g \circ f^{\leftarrow} \leqslant h \circ f^{\leftarrow}$. ∎

Since $\overline{G}$, $\overline{G}_n$, $\overline{F}_i$ and $\overline{F}$ are of the same order, so are their truncated means $\int_t^\infty x^k\,dK(x)$ for $K = \overline{G}, \overline{G}_n, \overline{F}_i$ and $\overline{F}$. Order is related to asymptotic behavior, and because we will need to have estimates that are uniform over n, we need a quantitative statement comparing truncated means. The following will do for our purposes and may be useful in other contexts.

LEMMA 6.3.3. *Let $\overline{F}$ be a normalized regularly varying function of index $-\alpha$. Let δ be a positive number. There exists t_4 such that for any $0 < u \leqslant v$, any $t \geqslant t_4$, any integer $k < \alpha$,*
$$\int_{tv}^\infty x^k\,dM_u F(x) \leqslant \frac{2\alpha}{\alpha - k} v^k \left(\frac{u}{v}\right)^{\alpha-\delta} t^k \overline{F}(t).$$

PROOF. A change of variable yields
$$\int_{tv}^\infty x^k\,dM_u F(x) = u^k \int_{tv/u}^\infty x^k\,dF(x).$$

By Karamata's theorem for Stieltjes integrals (Bingham et al., 1989, Theorem 1.6.5), there exists a t_4' such that for t more than t_4'

$$\int_t^\infty x^k \, dF(x) \leqslant \frac{2\alpha}{\alpha - k} t^k \overline{F}(t).$$

This t_4' may be taken independent of k since there is only a finite number of integers less than α. Since $v/u \geqslant 1$, this implies that for t at least t_4',

$$\int_{tv}^\infty x^k \, dM_u F(x) \leqslant \frac{2\alpha}{\alpha - k} v^k t^k \overline{F}(tv/u).$$

By Lemma 5.2.4, if $t \geqslant t_2$, we have $\overline{F}(tv/u) \leqslant (u/v)^{\alpha - \delta} \overline{F}(t)$. Take $t_4 = t_2 \vee t_4'$ to conclude. ∎

LEMMA 6.3.4. *Let $\overline{F}$ be a normalized regularly varying function with index $-\alpha$. Let ϵ be a positive number. There exists t_5 such that for any $t \geqslant t_5$, any $n \geqslant 1$, any sequence $(c_n)_{n \geqslant 1}$ with $C_\rho \leqslant 1$, and any $k < \alpha$,*

$$\int_t^\infty x^k \, dG_n(x) \leqslant \int_t^\infty x^k \, dG(x) \leqslant \frac{2\alpha}{\alpha - k} C_{\alpha(1-\rho) - \epsilon} t^k \overline{F}(t).$$

PROOF. Combining Lemmas 6.3.1 and 6.3.2, we have

$$\int_t^\infty x^k \, dG_n(x) \leqslant \int_t^\infty x^k \, dG(x) \leqslant \sum_{n \geqslant 1} \int_t^\infty x^k \, dM_{c_n^{1-\rho}} F(x),$$

where we used that C_ρ is at most 1 to obtain the last inequality. Since C_ρ is at most 1, so are all the c_n's. We apply Lemma 6.3.3 with $\delta = \epsilon/(1-\rho)$ to obtain

$$\int_t^\infty x^k \, dM_{c_n^{1-\rho}} F(x) \leqslant \frac{2\alpha}{\alpha - k} c_n^{(1-\rho)(\alpha - \delta)} t^k \overline{F}(t)$$

for t at least t_4. ∎

We will also use the following bound on the moments of G.

LEMMA 6.3.5. *Assume C_ρ is at most 1. For any s at least 1,*

$$\mu_{G,s} \leqslant C_{s(1-\rho)} \mu_{F,s}.$$

PROOF. Write

$$\mu_{G,s} = s \int_0^\infty x^{s-1} \overline{G}(x) \, dx.$$

Apply Lemma 6.3.1 to obtain that the moment of order s of G is at most

$$\sum_{n \geqslant 1} \int_0^\infty s x^{s-1} M_{c_n^{1-\rho}} \overline{F}(x) \, dx = \sum_{n \geqslant 1} \int_0^\infty (c_n^{1-\rho})^s x^s \, dF(x)$$

$$= C_{s(1-\rho)} \mu_{F,s}. \qquad \blacksquare$$

6.4. The fundamental estimate.

The origin of our asymptotic expansions is the following estimate, whose proof is almost deceptively simple.

THEOREM 6.4.1. *Let m be a positive integer and let r be in $[0,1)$. Furthermore, let K be a distribution function on the nonnegative half-line, whose m-th moment is finite. If h is smoothly varying of order $m+r$, then*

$$|(T_{K,\eta} - L_{K,m})h|(t) \leqslant \sum_{0 \leqslant j \leqslant m} \frac{|h^{(j)}(t)|}{j!} \int_{\eta t}^{\infty} x^j \, \mathrm{d}K(x)$$
$$+ \frac{|h^{(m)}(t)|}{t^r m!} \int_0^{\eta t} \overline{\Delta}^r_{t,x/t}(h^{(m)}) x^{m+r} \, \mathrm{d}K(x).$$

PROOF. The expression of $T_{K,\eta}$ and Proposition 5.2.5 show that

$$\left| T_{K,\eta} h(t) - \int_0^{\eta t} \sum_{0 \leqslant j \leqslant m} (-1)^j \frac{x^j}{j!} h^{(j)}(t) \, \mathrm{d}K(x) \right|$$
$$\leqslant \frac{|h^{(m)}(t)|}{t^r m!} \int_0^{\eta t} \overline{\Delta}^r_{t,x/t}(h^{(m)}) x^{m+r} \, \mathrm{d}K(x). \qquad (6.4.1)$$

Since

$$\int_0^{\eta t} (-1)^j \frac{x^j}{j!} h^{(j)}(t) \, \mathrm{d}K(x) = \frac{(-1)^j}{j!} \Big(\mu_{K,j} - \int_{\eta t}^{\infty} x^j \, \mathrm{d}K(x) \Big) \mathrm{D}^j h(t),$$

the result follows from the triangle inequality. ∎

We could also bound (6.4.1) by the simpler estimate

$$t^{-r} |h^{(m)}(t)| \overline{\Delta}^r_{t,\eta}(h^{(m)}) \mu_{K,m+r}.$$

Unfortunately, this is not good enough for our purposes.

Let us now explain why the estimate in the previous Theorem is useful. When h is smoothly varying and has index $-\beta$, the term of smallest order in $L_{K,m}h$ is given by $h^{(m)}$, which is in $RV_{-\beta-m}$. If $\overline{K}$ is regularly varying with index $-\alpha$, Karamata's theorem for Stieltjes integrals (Bingham et al., 1989, Theorem 1.6.5) shows that the function $t \mapsto \int_{\eta t}^{\infty} x^i \, \mathrm{d}K(x)$ is in $RV_{-\alpha+i}$. Thus, $h^{(i)}(t) \int_{\eta t}^{\infty} x^i \, \mathrm{d}K(x)$ is regularly varying with index $-\alpha - \beta$; this index is smaller than $-\beta - m$ if $m < \alpha$. Thus, Theorem 6.4.1 proves that when h is smoothly varying, $T_{K,\eta}h \sim L_{K,m}h$, and, moreover, gives an estimate of the error term of this asymptotic expansion.

6.5. Basic lemmas.

The goal of this subsection is to give estimates for the first three terms of the right hand side of (6.2.1).

Our first lemma takes care of the term $(T_{G_{n-1},1-d_n} - L_{G_{n-1},m})\overline{F}_n^{(k)}$. It asserts that it is of smaller order than $t^{-m-k}\overline{F}(t)$, uniformly in n and in some sequences $(c_n)_{n\geqslant 1}$. We write $v(t)$ for a function such that $0 \leqslant v(t) \leqslant t$ for all nonnegative t, and $\lim_{t\to\infty} v(t) = +\infty$ while $\lim_{t\to\infty} v(t)/t = 0$. We assume moreover that $v(t)/t$ is ultimately nonincreasing. For instance, $v(t) = t^\kappa$ with $0 < \kappa < 1$ will do.

LEMMA 6.5.1. *Let ϵ be a positive number. Assume that $\overline{F}$ is in $SR_{-\alpha,m+r+k}$ with m an integer, r in $[0,1)$ and $m+r$ smaller than α. There exist M and t_6 depending only on F, m, k and ϵ such that whenever $C_\rho \vee C_{(1-\rho)(m+r)} \leqslant 1$, and whenever $d_n \leqslant 1/2$, for any t at least t_6,*

$$|(T_{G_{n-1},1-d_n} - L_{G_{n-1},m})\overline{F}_n^{(k)}|_{k+m,t}$$
$$\leqslant Mc_n^{\alpha-(\alpha+k+m)\rho-\epsilon}\left(\overline{\Delta}_{t,v(t)/t}^r(\overline{F}^{(k+m)}) + v(t)^{m+r}\overline{F}\circ v(t)\right).$$

PROOF. Applying Theorem 6.4.1, we obtain for t at least 1,

$$|(T_{G_{n-1},1-d_n} - L_{G_{n-1},m})\overline{F}_n^{(k)}|(t)$$
$$\leqslant \sum_{0\leqslant j\leqslant m} |\overline{F}_n^{(k+j)}(t)| \int_{t(1-d_n)}^\infty x^j \, dG_{n-1}(x)$$
$$+ |\overline{F}_n^{(k+m)}(t)| \int_0^{t(1-d_n)} \overline{\Delta}_{t,x/t}^r(\overline{F}_n^{(k+m)}) x^{m+r} \, dG_{n-1}(x). \quad (6.5.1)$$

Since the c_i's are all at most 1, it suffices to prove the result for ϵ small enough. Thus, we can assume that $m+r$ is at most $\alpha-\epsilon$. Let $\delta = \epsilon/(2-\rho)$. We can assume that ϵ is small enough so that $(1-\rho)\alpha - \delta$ is at least $(1-\rho)(m+r)$. Then, the sum $C_{\alpha(1-\rho)-\delta}$ is at most $C_{(1-\rho)(m+r)}$, hence at most 1. Combining Lemmas 5.2.3 and 6.3.4, we see that for $t(1-d_n) \geqslant t_1 \vee t_5$,

$$|\overline{F}_n^{(k+j)}(t)| \int_{t(1-d_n)}^\infty x^j \, dG_{n-1} \leqslant Mc_n^{\alpha-\delta} t^{-k-j}\overline{F}(t) t^j \overline{F}(t)$$
$$\leqslant Mc_n^{\alpha-\delta} t^{-k}\overline{F}(t)^2.$$

Applying Lemmas 5.2.3, 5.2.6, 6.3.1 and 6.3.2, we also have, for t at least t_1 and t_2,

$$|\overline{F}_n^{(k+m)}(t)| \int_0^{t(1-d_n)} \overline{\Delta}_{t,x/t}^r(\overline{F}_n^{(k+m)}) x^{m+r} \, dG_{n-1}(x)$$
$$\leqslant Mc_n^{\alpha-\delta} t^{-k-m}\overline{F}(t) \int_0^{t(1-d_n)} \overline{\Delta}_{t/c_n,x/t}^r(\overline{F}^{(k+m)}) x^{m+r} \, dG_{n-1}(x).$$

We split the integral in this upper bound as one for x between 0 and $v(t)$, and one between $v(t)$ and $t(1-d_n)$. We then have the easy bound

$$\int_0^{v(t)} \overline{\Delta}^r_{t/c_n,x/t}(\overline{F}^{(k+m)}) x^{r+m} \, \mathrm{d}G_{n-1}(x) \leqslant \overline{\Delta}^r_{t,v(t)/t}(\overline{F}^{(k+m)}) \mu_{G,m+r}.$$

We use Lemma 6.3.5 to bound $\mu_{G,m+r}$, hence $\mu_{G_{n-1},m+r}$, by $C_{(m+r)(1-\rho)}\mu_{F,m+r}$, which is then at most $\mu_{F,m+r}$ under the assumptions of the lemma.

To bound the integral for x between $v(t)$ and $t(1-d_n)$, we first bound $\overline{\Delta}^r_{t/c_n,x/t}(\overline{F}^{(k+m)})$ by $\overline{\Delta}^r_{t/c_n,1-d_n}(\overline{F}^{(k+m)})$. The latter is at most

$$\overline{\Delta}^r_{t/c_n,1/2}(\overline{F}^{(k+m)}) + \sup_{s \geqslant t/c_n} \sup_{1/2 < |y| \leqslant 1-d_n} y^{-r}\left(\left|\frac{\overline{F}^{(k+m)}(s(1-y))}{\overline{F}^{(k+m)}(s)}\right| + 1\right).$$

If s is at least t/c_n and y at most $1-d_n$, then $s(1-y)$ is at least td_n/c_n, and therefore at least t. Thus, by the standard Potter bounds, there exists t'_6 such that for s at least t'_6,

$$\left|\frac{\overline{F}^{(k+m)}(s(1-y))}{\overline{F}^{(k+m)}(s)}\right| \leqslant 2\big((1-y)^{-\alpha-k-m-\delta} \vee (1-y)^{-\alpha-k-m+\delta}\big),$$

which is at most $2d_n^{-\alpha-k-m-\delta}$. Consequently,

$$\int_{v(t)}^{t(1-d_n)} \overline{\Delta}^r_{t/c_n,x/t}(\overline{F}^{(k+m)}) x^{m+r} \, \mathrm{d}G_{n-1}(x)$$

$$\leqslant M\big(\overline{\Delta}^r_{t/c_n,1/2}(\overline{F}^{(k+m)}) + d_n^{-\alpha-k-m-\delta} + 1\big)\int_{v(t)}^{\infty} x^{m+r} \, \mathrm{d}G_{n-1}(x).$$

Using Lemma 6.3.2 and since d_n is smaller than $1/2$, we can simplify this upper bound into

$$Md_n^{-\alpha-k-m-\epsilon}\int_{v(t)}^{\infty} x^{m+r} \, \mathrm{d}G(x).$$

Applying Lemma 6.3.4, for $v(t)$ at least t_5, this last quantity is at most

$$Md_n^{-\alpha-k-m-\delta} v(t)^{m+r}\overline{F}\circ v(t).$$

It follows that

$$\int_0^{t(1-d_n)} \overline{\Delta}^r_{t/c_n,x/t}(\overline{F}^{(k+m)}) x^{m+r} \, \mathrm{d}G(x)$$

$$\leqslant M\overline{\Delta}^r_{t,v(t)/t}(\overline{F}^{(k+m)}) + Md_n^{-\alpha-k-m-\delta} v(t)^{m+r}\overline{F}\circ v(t).$$

Combined with the other bounds, we obtain that the right hand side of (6.5.1) is at most

$$Mc_n^{\alpha-\delta}t^{-k-m}\overline{F}(t)\Big(t^m\overline{F}(t) + \overline{\Delta}^r_{t,v(t)/t}(\overline{F}^{(k+m)}) + d_n^{-\alpha-k-m-\delta}v(t)^{m+r}\overline{F}\circ v(t)\Big).$$

Since $m+r$ is smaller than α, the function $s \mapsto s^{m+r}\overline{F}(s)$ is ultimately nonincreasing (Bingham et al., 1989, Theorem 1.5.3). Thus, for t large enough, $t^m\overline{F}(t)$ is smaller than $2v(t)^{m+r}\overline{F} \circ v(t)$. The proof is completed upon noting that $d_n^{-\alpha-k-m-\delta}$ is at least 1. ∎

Our next lemma shows that when d_n is small, that is n is large, then T_{F_n,d_n} is very close to being a contraction.

LEMMA 6.5.2. *There exists some positive numbers M and t_7 such that for any t at least t_7, any positive n, any k, any d_n smaller than $1/2$, any function h,*

$$|T_{F_n,d_n}h|_{k,t} \leqslant |h|_{k,t(1-d_n)}(1+Md_n).$$

PROOF. A change of variable shows that

$$T_{F_n,d_n}h(t) = \int_0^{td_n/c_n} h(t-c_n x)\,\mathrm{d}F(x).$$

If x is at most td_n/c_n, then $t-c_n x$ is at least $t(1-d_n)$. Consequently,

$$|h(t-c_n x)| \leqslant |h|_{k,t(1-d_n)}\frac{\overline{F}(t-c_n x)}{(t-c_n x)^k} \leqslant |h|_{k,t(1-d_n)}\frac{\overline{F}\bigl(t(1-d_n)\bigr)}{t^k(1-d_n)^k}.$$

This implies

$$\frac{t^k|T_{F_n,d_n}h(t)|}{\overline{F}(t)} \leqslant |h|_{k,t(1-d_n)}\frac{\overline{F}\bigl(t(1-d_n)\bigr)}{\overline{F}(t)}(1-d_n)^{-k}.$$

Since d_n is at most $1/2$, we can apply Lemma 5.2.4, and then bound $(1-d_n)^{-\alpha-k-\delta}$ by $1+Md_n$. ∎

REMARK. In proving Lemma 6.5.2, our improved Potter bound is crucial. With the usual one we could only get $A(1+Md_n)$ for some A greater than 1. This is not good enough for the inductions to come. Extensive attempts to prove our theorem under weaker assumptions suggest that the behavior of the function $c(\cdot)$ in the Karamata representation (see Bingham, Goldie and Teugels, 1989, Theorem 1.3.1) of $\overline{F}$ cannot be arbitrary for our asymptotic expansion to hold. In particular, the asymptotic behavior of $c(\cdot)$ plays a key role in the asymptotic expansion. This is already the case for a two terms expansion where the second term involves derivative; one should indeed remark that the absolute continuity of the function $c(\cdot)$ is equivalent to that of $\overline{F}$.

In particular, in connection with the second term in (6.2.1), we obtain the following estimate, by considering $h = \overline{G}_{n-1}^{(k)} - \mathcal{A}_m\overline{G}_{n-1}^{(k)}$ in Lemma 6.5.2.

COROLLARY 6.5.3. *There exist positive numbers M and t_8 such that for any t at least t_8 and any sequence $(c_n)_{n\geqslant 1}$ with $(d_n)_{n\geqslant 1}$ uniformly bounded by $1/2$, any k and any n at least 2,*

$$|T_{F_n,d_n}(\overline{G}_{n-1}^{(k)} - \mathcal{A}_m \overline{G}_{n-1}^{(k)})|_{k+m,t} \leqslant |\overline{G}_{n-1}^{(k)} - \mathcal{A}_m \overline{G}_{n-1}^{(k)}|_{k+m,t(1-d_n)}(1 + Md_n).$$

This estimate shows the need to introduce the family of operators $T_{K,\eta}$ and not use $T_{K,1/2}$ as in Barbe and McCormick (2005). Indeed, when doing the induction later, we see that the value of t in the norm of the right hand side drops by a factor $1 - d_n$. When applying this bound inductively, we will obtain something like

$$|\overline{G}_2^{(k)} - \mathcal{A}\overline{G}_2^{(k)}|_{k+m,t\Pi_{1\leqslant i\leqslant n}(1-d_i)}.$$

If $\prod_{i\geqslant 1}(1 - d_i) = 0$, we will not be able to permute the limits as n tends to infinity and t tends to infinity. It is therefore essential that the series $\sum_{n\geqslant 1} d_n$ converges.

Our last lemma in this section will handle the third term in the right hand side of (6.2.1).

LEMMA 6.5.4. *Let ϵ be a positive number at most $\alpha - 1$. Assume that F is in $SR_{-\alpha,m+k+r}$. There exist positive M and t_9 such that for any sequence $(c_n)_{n\geqslant 1}$ with C_ρ at most 1, any t at least t_9 and any n at least 2,*

$$\left| T_{F_n,d_n} \mathcal{A}_m \overline{G}_{n-1}^{(k)} - \sum_{1\leqslant i\leqslant n-1} L_{G_{n\setminus i}} \overline{F}_i^{(k)} \right|_{k+m,t}$$
$$\leqslant M\left(c_n^{\alpha(1-\rho)-\epsilon} + c_n^r \overline{\Delta}_{t,d_n}^r(\overline{F}^{(k+m)})\right) v(t)^{m+r}\overline{F}\circ v(t) + Mc_n^r \overline{\Delta}_{t,c_n v(t)/t}^r(\overline{F}^{(k+m)}).$$

The proof of this lemma requires an auxiliary result which we state as a claim. It compares $T_{K,\eta} L_H$ with $L_{K\star H}$ (cf. subsection 6.2).

CLAIM. *For any nonnegative t,*

$$|T_{K,\eta} L_H h - L_{K\star H} h|(t) \leqslant \sum_{0\leqslant s\leqslant m} \frac{|h^{(s)}(t)|}{s!} \sum_{0\leqslant j\leqslant s} \binom{s}{j} \mu_{H,j} \int_{\eta t}^\infty x^{s-j}\,\mathrm{d}K(x)$$
$$+ \frac{|h^{(m)}(t)|}{t^r m!} \sum_{0\leqslant j\leqslant m} \binom{m}{j} \mu_{H,j} \int_0^{\eta t} \overline{\Delta}_{t,x/t}^r(h^{(m)}) x^{m+r-j}\,\mathrm{d}K(x).$$

PROOF. By linearity of $T_{K,\eta}$ we have

$$T_{K,\eta} L_H = \sum_{0\leqslant j\leqslant m} \frac{(-1)^j}{j!} \mu_{H,j} T_{K,\eta} \mathrm{D}^j.$$

Applying Theorem 6.4.1 to $h^{(j)}$ in the form of a bound for $|T_{K,\eta}h^{(j)} - L_{K,m-j}h^{(j)}|$, we obtain

$$\left| T_{K,\eta}L_{H,m}h - \sum_{0\leqslant j\leqslant m} \frac{(-1)^j}{j!} \mu_{H,j} L_{K,m-j} h^{(j)} \right|(t)$$

$$\leqslant \sum_{0\leqslant j\leqslant m} \frac{\mu_{H,j}}{j!} \bigg(\sum_{0\leqslant l\leqslant m-j} \frac{|h^{(j+l)}(t)|}{l!} \int_{\eta t}^{\infty} x^l \, \mathrm{d}K(x)$$

$$+ \frac{|h^{(m)}(t)|}{t^r(m-j)!} \int_0^{\eta t} \overline{\Delta}^r_{t,x/t}(h^{(m)}) x^{m-j+r} \, \mathrm{d}K(x) \bigg).$$

Using Lemma 5.1.3, the left hand side is the absolute value of $(T_{K,\eta}L_H - L_{K\star H})h$ evaluated at t. Setting $s = l + j$, the right hand side is at most

$$\sum_{s,j} \mathbb{1}\{j \leqslant s \leqslant m\} \frac{|h^{(s)}(t)|}{s!} \binom{s}{j} \mu_{H,j} \int_{\eta t}^{\infty} x^{s-j} \, \mathrm{d}K(x)$$

$$+ \frac{|h^{(m)}(t)|}{t^r m!} \bigg(\sum_{0\leqslant j\leqslant m} \binom{m}{j} \mu_{H,j} \int_0^{\eta t} \overline{\Delta}^r_{t,x/t}(h^{(m)}) x^{m-j+r} \, \mathrm{d}K(x) \bigg).$$

This is our claim. ∎

PROOF (of Lemma 6.5.4). First the triangle inequality implies the pointwise inequality

$$\left| T_{F_n,d_n} \mathcal{A}\overline{G}^{(k)}_{n-1} - \sum_{1\leqslant i\leqslant n-1} L_{G_{n\setminus i}} \overline{F}^{(k)}_i \right|$$
$$\leqslant \sum_{1\leqslant i\leqslant n-1} |T_{F_n,d_n} L_{G_{n-1\setminus i}} \overline{F}^{(k)}_i - L_{G_{n\setminus i}} \overline{F}^{(k)}_i|. \quad (6.5.2)$$

Since $F_n \star (G_{n-1\setminus i}) = G_{n\setminus i}$, the claim yields for t at least 1,

$$|T_{F_n,d_n} L_{G_{n-1\setminus i}} \overline{F}^{(k)}_i - L_{G_{n\setminus i}} \overline{F}^{(k)}_i|(t)$$
$$\leqslant \sum_{0\leqslant s\leqslant m} \frac{|\overline{F}^{(k+s)}_i(t)|}{s!} \sum_{0\leqslant j\leqslant s} \binom{s}{j} \mu_{G_{n-1\setminus i},j} \int_{td_n}^{\infty} x^{s-j} \, \mathrm{d}F_n(x)$$
$$+ \frac{|\overline{F}^{(k+m)}_i(t)|}{m!} \sum_{0\leqslant j\leqslant m} \binom{m}{j} \mu_{G_{n-1\setminus i},j} \int_0^{d_n t} \overline{\Delta}^r_{t,x/t}(\overline{F}^{(k+m)}_i) x^{m+r-j} \, \mathrm{d}F_n(x). \quad (6.5.3)$$

Since we assume C_ρ to be at most 1, all the c_n's are at most 1. From Lemma 5.2.3, we deduce that for t at least t_1,

$$|\overline{F}^{(k+s)}_i(t)| \leqslant M c_i^{\alpha-\epsilon} t^{-k-s} \overline{F}(t).$$

Moreover, $\mu_{G_{n-1\setminus i},j} \leqslant \mu_{G,j}$. But $\mu_{G,j}$ is a fixed polynomial in the C_l's and $\mu_{F,l}$'s, $1 \leqslant l \leqslant j$. Since all these C_l's are at most C_ρ, we conclude that $\mu_{G_{n-1\setminus i},j}$ is at most some fixed constant M; this constant does not depend on n and sequences $(c_n)_{n\geqslant 1}$ with $C_\rho \leqslant 1$.

Lemma 6.3.3 implies that for any t at least t_4,
$$\int_{td_n}^\infty x^{s-j}\,\mathrm{d}F_n(x) \leqslant M d_n^{s-j} c_n^{\alpha(1-\rho)-\epsilon} t^{s-j}\overline{F}(t).$$

It follows that the sum in the first term of (6.5.3) is at most
$$\sum_{0\leqslant s\leqslant m} Mc_i^{\alpha-\epsilon}t^{-k-s}\overline{F}(t) \sum_{0\leqslant j\leqslant s} d_n^{s-j}c_n^{\alpha(1-\rho)-\epsilon}t^{s-j}\overline{F}(t) \leqslant Mt^{-k}\overline{F}(t)^2 c_i^{\alpha-\epsilon}c_n^{\alpha(1-\rho)-\epsilon}.$$

Similarly, for the second term in (6.5.3), we use the same bounds and Lemma 5.2.6 to obtain that it is at most
$$Mc_i^{\alpha-\epsilon}t^{-k-m}\overline{F}(t)\max_{0\leqslant j\leqslant m}\int_0^{d_n t}\overline{\Delta}_{t/c_i,x/t}^r(\overline{F}^{(k+m)})x^{m+r-j}\,\mathrm{d}F_n(x).$$

Using a change of variables, we rewrite the integral terms in this bound as
$$c_n^{m+r-j}\int_0^{td_n/c_n}\overline{\Delta}_{t/c_i,c_n x/t}^r(\overline{F}^{(k+m)})x^{m+r-j}\,\mathrm{d}F(x).$$

As in the proof of Lemma 6.5.1, we split the integral in this bound into one for x between 0 and $v(t)$, and one between $v(t)$ and td_n/c_n. We have
$$\int_0^{v(t)}\overline{\Delta}_{t/c_i,c_n x/t}^r(\overline{F}^{(k+m)})x^{m+r-j}\,\mathrm{d}F(x) \leqslant \overline{\Delta}_{t,c_n v(t)/t}^r(\overline{F}^{(k+m)})\mu_{F,m+r-j},$$

and using Lemma 6.3.3 (with $u=v=1$ in it), for $v(t)$ at least t_1,
$$\int_{v(t)}^{td_n/c_n}\overline{\Delta}_{t/c_i,c_n x/t}^r(\overline{F}^{(k+m)})x^{m+r-j}\,\mathrm{d}F(x) \leqslant M\overline{\Delta}_{t/c_i,d_n}^r(\overline{F}^{(k+m)})v(t)^{m+r-j}\overline{F}\circ v(t).$$

We conclude that for t large enough (6.5.3) is at most
$$Mt^{-k-m}\overline{F}(t)c_i^{\alpha-\epsilon}\Big(t^m\overline{F}(t)c_n^{\alpha(1-\rho)-\epsilon} + c_n^r\overline{\Delta}_{t,c_n v(t)/t}^r(\overline{F}^{(k+m)})$$
$$+ c_n^r\overline{\Delta}_{t,d_n}^r(\overline{F}^{(k+m)})v(t)^{m+r}\overline{F}\circ v(t)\Big).$$

Since $m+r$ is smaller than α, the function $s\mapsto s^{m+r}\overline{F}(s)$ is asymptotically equivalent to a nonincreasing function (Bingham et al., Theorem 1.5.3). Hence, for $v(t)$ large enough, $t^m\overline{F}(t)$ is at most $2v(t)^{m+r}\overline{F}\circ v(t)$. Therefore, (6.5.3) is at most
$$Mt^{-k-m}\overline{F}(t)c_i^{\alpha-\epsilon}\Big(\big(c_n^{\alpha(1-\rho)-\epsilon} + c_n^r\overline{\Delta}_{t,d_n}^r(\overline{F}^{(k+m)})\big)v(t)^{m+r}\overline{F}\circ v(t)$$
$$+ c_n^r\overline{\Delta}_{t,c_n v(t)/t}^r(\overline{F}^{(k+m)})\Big).$$

We then use (6.5.2) to finish. ∎

6.6. Inductions.

Define
$$\gamma_{m,n}^{(k)}(t) = |\overline{G}_n^{(k)} - \mathcal{A}_m \overline{G}_n^{(k)}|_{k+m,t}.$$

Our proof of Theorem 2.5.1 consists in bounding $\gamma_{m,n}^{(k)}(t)$ uniformly in n. This is done by induction on n, and we first need to settle the case $n=2$.

LEMMA 6.6.1. *Take ρ less than $\alpha/(\alpha+k)$. There exists a function η_0 with limit 0 at infinity, such that whenever d_1 and d_2 are at most $1/2$ and $k+m < \omega$, for any t nonnegative,*
$$\gamma_{m,2}^{(k)}(t) \leqslant \eta_0(t).$$

PROOF. Proposition 6.1.1 shows that
$$\overline{G}_2 = T_{F_1,1-d_2}\overline{F}_2 + T_{F_2,d_2}\overline{F}_1 + M_{1/(1-d_2)}\overline{F}_1 M_{1/d_2}\overline{F}_2, \qquad (6.6.1)$$

while for k at least 1,
$$\overline{G}_2^{(k)} = T_{F_1,1-d_2}\overline{F}_2^{(k)} + T_{F_2,d_2}\overline{F}_1^{(k)} - \sum_{1\leqslant j\leqslant k-1} M_{1/(1-d_2)}\overline{F}_1^{(j)} M_{1/d_2}\overline{F}_2^{(k-j)}. \qquad (6.6.2)$$

Let r be a nonnegative number less than 1 and $\omega-k-m$. We use a slight modification of Lemma 6.5.1 with n equal 2 to obtain
$$|(T_{F_1,1-d_2} - L_{F_1,m})\overline{F}_2^{(k)}|_{k+m,t}$$
$$\leqslant Mc_2^{\alpha-\rho(\alpha+k+m)-\epsilon}\big(\overline{\Delta}_{t,v(t)/t}^r(\overline{F}^{(k+m)}) + v(t)^{m+r}\overline{F}\circ v(t)\big).$$

Permuting F_1 and F_2 yields an upper bound for the error committed in approximating $T_{F_2,d_2}\overline{F}_1^{(k)}$ by $L_{F_2,m}\overline{F}_1^{(k)}$.
Combined with (6.6.2), this proves the lemma when k is 1.
When k vanishes, since m is smaller than α,
$$M_{1/d_2}\overline{F}_2(t) M_{1/(1-d_2)}\overline{F}_1(t) = o\big(t^{-m}\overline{F}(t)\big)$$

as t tends to infinity. Since $M_{1/d_2}\overline{F}_2 = M_{c_2^{1-\rho}}\overline{F}$ this is uniform in d_1 and d_2 at most $1/2$. Combined with (6.6.1), this proves the result for k vanishing.
When k is larger than 1, let ϵ be less than $\alpha(1-\rho) - k\rho$. We use Lemma 5.2.3 to obtain
$$|M_{1/d_2}\overline{F}_2^{(i)} M_{1/(1-d_2)}\overline{F}_1^{(k-i)}|(t) = \big|c_2^{-i}\overline{F}^{(i)}(td_2/c_2)c_1^{i-k}\overline{F}^{(k-i)}\big(t(1-d_2)/c_1\big)\big|$$
$$\leqslant Mt^{-i}c_2^{\alpha(1-\rho)-i\rho-\epsilon}\overline{F}(t)t^{-k+i}c_1^{\alpha-\epsilon}\overline{F}(t)$$

which is at most $Mt^{-k}\overline{F}(t)^2$. We conclude as before. ∎

The next step is to show that $\gamma_{m,n}^{(k)}$ can be bounded by induction. For this purpose, we need another bound that will control the sum involving the multiplication operators in (6.2.1).

LEMMA 6.6.2. *Let k be at least 2. Let ϵ be a positive real number. Assume that all d_i's are smaller than $1/2$. There exist t_9 and M such that for any t at least t_9, any positive integers j, k with $j \leqslant k-1$,*

$$\sum_{i \geqslant 2} |M_{1/d_i}\overline{F}_i^{(j)} M_{1/(1-d_i)}\overline{G}_{i-1}^{(k-j)}|_{k+m,t}$$
$$\leqslant Mt^m \overline{F}(t)\big(C_{\alpha-\epsilon/(1-\rho)} + \sup_{i \geqslant 2} \gamma_{0,i}^{(k-j)}(t/2)\big)C_{\alpha-\rho(\alpha+j)-\epsilon}\,.$$

PROOF. Define $\delta = \epsilon/(1-\rho)$. Under our assumptions, all the c_i's are at most 1. Consequently, d_i/c_i is at least 1 and Lemma 5.2.3 (taking c to be 1 in that lemma) implies that for s at least t_1,

$$|M_{1/d_i}\overline{F}_i^{(j)}(s)| = c_i^{-j}|\overline{F}^{(j)}(sd_i/c_i)| \leqslant c_i^{-j}M(sd_i/c_i)^{-j}\overline{F}(sd_i/c_i)\,.$$

Applying Lemma 5.2.4, using again the fact that d_i/c_i is at least 1, we see that for s at least t_1 and t_2,

$$|M_{1/d_i}\overline{F}_i^{(j)}(s)| \leqslant Mc_i^{-j+(j+\alpha-\delta)(1-\rho)}s^{-j}\overline{F}(s)\,.$$

Since $\mathcal{A}_0 \overline{G}_i^{(k-j)} = \sum_{1 \leqslant l \leqslant i} \overline{F}_l^{(k-j)}$, the triangle inequality yields

$$|\overline{G}_i^{(k-j)}(s)| \leqslant \sum_{1 \leqslant l \leqslant i} |\overline{F}_l^{(k-j)}(s)| + \gamma_{0,i}^{(k-j)}(s)s^{-k+j}\overline{F}(s)\,.$$

Using again Lemma 5.2.3 and the fact that the c_i's are at most 1, we see that for s at least t_1,

$$|\overline{G}_i^{(k-j)}(s)| \leqslant M \sum_{1 \leqslant l \leqslant i} c_l^{\alpha-\delta}s^{-k+j}\overline{F}(s) + \gamma_{0,i}^{(k-j)}(s)s^{-k+j}\overline{F}(s)$$
$$\leqslant Ms^{-k+j}\overline{F}(s)\big(C_{\alpha-\delta} + \sup_{n \geqslant 2}\gamma_{0,n}^{(k-j)}(s)\big)\,.$$

Consequently, since d_i is at most $1/2$, we obtain that for t large enough,

$$\sum_{i \geqslant 2} \sup_{s \geqslant t} s^{k+m}|M_{1/d_i}\overline{F}_i^{(j)}(s)M_{1/(1-d_i)}\overline{G}_{i-1}^{(k-j)}(s)|/\overline{F}(s)$$
$$\leqslant M \sup_{s \geqslant t} s^m \overline{F}(s/2)\big(C_{\alpha-\delta} + \sup_{n \geqslant 2}\gamma_{0,n}^{(k-j)}(s/2)\big)C_{(j+\alpha-\delta)(1-\rho)-j}\,,$$

the factor $1/2$ in $s/2$ coming from the multiplication operator acting on $\overline{G}_{i-1}^{(k-1)}$ and our assumption that d_i is at most $1/2$. One more application of Potter's bound allows us to replace $\overline{F}(s/2)$ by $\overline{F}(s)$ in the upper bound. Since $\gamma_{0,i}^{(k-j)}(\cdot)$

is nonincreasing and $s \mapsto s^m \overline{F}(s)$ is asymptotically equivalent to an ultimately nonincreasing function (Bingham et al., 1989, Theorem 1.5.3), we can remove the supremum in s in the above bound (and, of course, we increase M in doing that). ∎

Our next lemma shows that $\gamma_{m,n}^{(k)}(\cdot)$ can be bounded by induction on k, whenever k is at least 2. We define

$$\eta_\sigma(t) = \bigl(v(t)/t\bigr)^\gamma + v(t)^\sigma \overline{F} \circ v(t).$$

LEMMA 6.6.3. *Let ϵ be a positive real number, and let k be a positive integer. There exist positive numbers M and t_{10} such that for any integer n at least 2 and any sequence $(c_n)_{n \geqslant 1}$ with $C_{\gamma\rho} \vee C_{\alpha-(\alpha+k+m)\rho-\epsilon} \leqslant 1$ and $d_n \leqslant 1/2$,*

$$\gamma_{m,n}^{(k)}(t) \leqslant M\Bigl(1 + \max_{1 \leqslant j \leqslant k-1} \sup_{i \geqslant 2} \gamma_{0,i}^{(k-j)}(t/M)\Bigr)\eta_m(t/M) + M\gamma_{m,2}^{(k)}(t/M).$$

In the statement of this lemma, we agree that $\max_{1 \leqslant j \leqslant 0}$ is 0.

PROOF. For any u and $\gamma - r$ positive, the inequality $\overline{\Delta}_{t,u}^r(h) \leqslant u^{\gamma-r}\overline{\Delta}_{t,u}^\gamma(h)$ holds. We will take r to be 0 in this inequality and apply it in the bounds of the previous subsection. Under the assumption of Theorem 2.5.1, we also have $\limsup_{t \to \infty} \overline{\Delta}_{t,1/2}^0(\overline{F}^{(k+m)})$ finite. Using (6.2.1), Lemma 6.5.1, Corollary 6.5.3 and Lemma 6.5.4, we obtain that there exists M and t_{10}' such that for any t at least t_{10}' and any n at least 3,

$$\begin{aligned}\gamma_{m,n}^{(k)}(t) \leqslant{}& Mc_n^{\alpha-(\alpha+k+m)\rho-\epsilon}\Bigl(\bigl(v(t)/t\bigr)^\gamma + v(t)^m \overline{F} \circ v(t)\Bigr) \\ & + \gamma_{m,n-1}^{(k)}\bigl(t(1-d_n)\bigr)(1+Md_n) \\ & + M(c_n^{\alpha(1-\rho)-\epsilon} + d_n^\gamma)v(t)^m \overline{F} \circ v(t) + Mc_n^\gamma\bigl(v(t)/t\bigr)^\gamma \\ & + \sum_{1 \leqslant j \leqslant k-1} |M_{1/d_n} \overline{F}_n^{(j)} M_{1/(1-d_n)} \overline{G}_{n-1}^{(k-j)}|_{k+m,t}\,.\end{aligned}$$

Collecting the terms and using that all the c_n's are at most 1,

$$\begin{aligned}\gamma_{m,n}^{(k)}(t) \leqslant{}& M(c_n^{\alpha-(\alpha+k+m)\rho-\epsilon} + c_n^{\gamma\rho})\eta_m(t) \\ & + \gamma_{m,n-1}^{(k)}\bigl(t(1-d_n)\bigr)(1+Md_n) \quad\quad\quad (6.6.3) \\ & + \sum_{1 \leqslant j \leqslant k-1} |M_{1/d_n} \overline{F}_n^{(j)} M_{1/(1-d_n)} \overline{G}_{n-1}^{(k-j)}|_{k+m,t}\,.\end{aligned}$$

Taking t_{10}' large enough, we can assume that $\eta_m(\cdot)$ is nonincreasing on $[t_{10}', \infty)$. Dropping the subscript m and superscript k temporarily, inequality (6.6.3) has the form

$$\gamma_n(t) \leqslant a_n(t) + \gamma_{n-1}\bigl(t(1-d_n)\bigr)(1+Md_n),$$

where $a_n(\cdot)$ is the nonnegative and nonincreasing function on $[t'_{10}, \infty)$ given by

$$a_n(t) = M(c_n^{\alpha-(\alpha+k+m)\rho-\epsilon} + c_n^{\rho\gamma})\eta_m(t)$$
$$+ \sum_{1\leqslant j\leqslant k-1} |M_{1/d_n}\overline{F}_n^{(j)}M_{1/(1-d_n)}\overline{G}_{n-1}^{(k-j)}|_{k+m,t}.$$

For $i+1$ at most n, define

$$A_{i,n} = \prod_{i+1\leqslant j\leqslant n}(1-d_j) \quad \text{and} \quad B_{i,n} = \prod_{i+1\leqslant j\leqslant n}(1+Md_j).$$

We also set $A_{n,n} = B_{n,n} = 1$. By induction, inequality (6.6.3) implies

$$\gamma_n(t) \leqslant \sum_{3\leqslant i\leqslant n} B_{i,n}a_i(tA_{i,n}) + B_{2,n}\gamma_2(tA_{2,n}).$$

Set $A = \prod_{j\geqslant 1}(1-d_j)$ and $B = \prod_{j\geqslant 1}(1+Md_j)$. Since all the $a_n(\cdot)$'s are nonnegative and nonincreasing, we have

$$\gamma_n(t) \leqslant B\sum_{i\geqslant 3} a_i(tA) + B\gamma_2(tA).$$

The inequality $\log(1-x) \geqslant -2x$ for x nonnegative and at most $1/2$ implies

$$A \geqslant \exp\Big(-2\sum_{j\geqslant 1} d_j\Big) = \exp(-2C_\rho),$$

while the inequality $\log(1+x) \leqslant x$ implies

$$B \leqslant \exp\Big(\sum_{j\geqslant 1} Md_j\Big) = \exp(MC_\rho).$$

Consequently,

$$\gamma_n(t) \leqslant e^{MC_\rho}\Big(\sum_{i\geqslant 3} a_i(te^{-2C_\rho}) + \gamma_2(te^{-2C_\rho})\Big).$$

We also have

$$\sum_{i\geqslant 3} a_i(s) \leqslant M(C_{\alpha-(\alpha+k+m)\rho-\epsilon} + C_{\rho\gamma})\eta_m(s)$$
$$+ \sum_{1\leqslant j\leqslant k-1}\sum_{i\geqslant 3} |M_{1/d_i}\overline{F}_i^{(j)}M_{1/(1-d_i)}\overline{G}_{i-1}^{(j-k)}|_{k+m,s}.$$

Applying Lemma 6.6.2, this bound is at most

$$M(C_{\alpha-(\alpha+k+m)\rho-\epsilon} + C_{\rho\gamma})\eta_m(s)$$
$$+ Ms^m\overline{F}(s)\big(C_{\alpha-\epsilon/(1-\rho)} + \max_{1\leqslant j\leqslant k-1}\sup_{i\geqslant 2}\gamma_{0,i}^{(k-j)}(s/2)\big)C_{\alpha-\rho(\alpha+k)-\epsilon}.$$

Since $s^m \overline{F}(s)$ is at most $\eta_m(s)$ for s at least 1, this bound yields

$$\gamma_n(t) \leqslant M\Big(C_{\alpha-(\alpha+k+m)\rho-\epsilon} + C_{\gamma\rho} + 1 + \max_{1\leqslant j\leqslant k-1} \sup_{i\geqslant 2} \gamma_{0,i}^{(k-j)}(te^{-2C_\rho}/2)\Big)$$
$$\times \eta_m(te^{-2C_\rho}) + e^{MC_\rho}\gamma_2(te^{-2C_\rho}).$$

To conclude the proof, if $C_{\rho\gamma}$ is at most 1, so are all the c_i's, implying that C_ρ is at most $C_{\rho\gamma}$. ∎

We need an analogue of Lemma 6.6.3 when k vanishes. For this purpose, we state the analogue of Lemma 6.6.2 in this case.

LEMMA 6.6.4. *Assume that C_ρ is at most 1 and that all d_i's are at most $1/2$. For t larger than some t_{11}, the following inequality holds:*

$$\sum_{i\geqslant 1} |\mathsf{M}_{1/d_i}\overline{F}_i \mathsf{M}_{1/(1-d_i)}\overline{G}_{i-1}|_{k+m,t} \leqslant C_{\alpha(1-\rho)-\epsilon}^2 t^m \overline{F}(t).$$

PROOF. Set $\delta = \epsilon/(1-\rho)$. Apply Lemma 5.2.4 to obtain that for s at least t_2,

$$\mathsf{M}_{1/d_i}\overline{F}_i(s) = \overline{F}(sd_i/c_i) \leqslant c_i^{(1-\rho)(\alpha-\delta)} \overline{F}(s).$$

Since d_i is at most $1/2$, Lemma 6.3.1 implies

$$\mathsf{M}_{1/(1-d_i)}\overline{G}_{i-1}(s) \leqslant \sum_{n\geqslant 1} \overline{F}(s/2c_n^{1-\rho}),$$

which, by Lemma 5.2.4 is at most $M\sum_{n\geqslant 1} c_n^{(1-\rho)(\alpha-\delta)} \overline{F}(s)$ for s large enough. This implies the conclusion. ∎

The next lemma is Theorem 2.5.1 stated in another way.

LEMMA 6.6.5. *Let ϵ be a positive number at most 1. There exists a function $\eta(\cdot)$ with limit 0 at infinity, such that, for any k and m with $k+m$ smaller than ω and m smaller than α, any n at least 2, and any t at least 1 say, any nonnegative sequence $(c_i)_{i\geqslant 1}$ with $\sup_{i\geqslant 1} d_i \leqslant 1/2$ and both $C_{\gamma\rho}$ and $C_{\alpha-(\alpha+k+m)\rho-\epsilon}$ at most 1,*

$$\gamma_{m,n}^{(k)}(t) \leqslant \eta(t).$$

PROOF. For $k = 1$, Lemma 6.6.3 implies

$$\gamma_{m,n}^{(1)}(t) \leqslant M\eta_m(t/M) + M\gamma_{m,2}^{(1)}(t/M).$$

The result follows then from the assumptions and Lemma 6.6.1.

By induction on k, Lemmas 6.6.3 and 6.6.2 imply the result for k at least 1.

When k vanishes, Lemma 6.6.4 and the same arguments as those in the proof of Lemma 6.6.3 show that

$$\gamma_{m,n}^{(0)}(t) \leqslant M(C_{(1-\rho)(\alpha-\epsilon)}+1)\eta_m(t/M) + M\gamma_{m,2}^{(0)}(t/M) + MC_{\alpha(1-\rho)-\epsilon}^2 t^m \overline{F}(t)$$

(compare with (6.6.3)). The result follows. ∎

6.7. Conclusion.

To obtain Theorem 2.5.1, it mostly remains to show that as n tends to infinity, $\gamma_{m,n}(t)$ converges to $|\overline{G}^{(k)} - \mathcal{A}_m \overline{G}^{(k)}|_{k+m,t}$. This is achieved in two steps: one consisting in proving that the sequence of approximations converges, the other one in proving that the sequence of functions $G_n^{(k)}$ converges to $G^{(k)}$.

LEMMA 6.7.1. *Assume that $C_{1-\rho}$ and $|c|_\infty$ are at most 1. Then there exists t_{12} and M such that for any t at least t_{12},*

$$|\mathcal{A}_m \overline{G}_n^{(k)} - \mathcal{A}_m \overline{G}^{(k)}|_{k,t} \leqslant M \sum_{i \geqslant n+1} c_i^{1-\rho}.$$

REMARK. For our purposes, it is enough that $\mathcal{A}_m \overline{G}_n^{(k)}$ converges pointwise to $\mathcal{A}_m \overline{G}^{(k)}$. However, in order to simulate properly tail behavior, it would be desirable to have convergence in $|\cdot|_{k+m,t}$ norm.

PROOF. From the definition of the approximation and since both $\mu_{G_n \setminus i, 0}$ and $\mu_{G \setminus i, 0}$ equal 1 for i at most n,

$$\mathcal{A}_m \overline{G}_n^{(k)} - \mathcal{A}_m \overline{G}^{(k)} = \sum_{0 \leqslant j \leqslant m} \frac{(-1)^j}{j!} \sum_{1 \leqslant i \leqslant n} (\mu_{G_n \setminus i, j} - \mu_{G \setminus i, j}) \mathrm{D}^{j+k} \overline{F}_i$$
$$- \sum_{0 \leqslant j \leqslant m} \frac{(-1)^j}{j!} \sum_{i \geqslant n+1} \mu_{G \setminus i, j} \mathrm{D}^{j+k} \overline{F}_i. \quad (6.7.1)$$

Write $Y = \sum_{\substack{1 \leqslant j \leqslant n \\ j \neq i}} c_j X_j$ and $R = \sum_{\substack{j \geqslant n+1 \\ j \neq i}} c_j X_j$. Then, for j at least 1,

$$\mu_{G \setminus i, j} - \mu_{G_n \setminus i, j} = E\big((Y+R)^j - Y^j\big) = \sum_{1 \leqslant l \leqslant j} \binom{j}{l} ER^l EY^{j-l}.$$

Applying Lemma 6.3.5, we see that for any l at least 1,

$$ER^l \leqslant \sum_{i \geqslant n+1} c_i^{l(1-\rho)} \mu_{F,l} \leqslant \sum_{i \geqslant n+1} c_i^{1-\rho} \mu_{F,l},$$

while for l less than j, by the same token and under the assumption of the lemma,

$$EY^{j-l} \leqslant \mu_{G,j-l} \leqslant C_{(j-l)(1-\rho)} \mu_{F,j-l} \leqslant \mu_{F,j-l}.$$

Consequently, for j at least 1,

$$0 \leqslant \mu_{G\setminus i,j} - \mu_{G_n\setminus i,j} \leqslant \sum_{1\leqslant l\leqslant j}\binom{j}{l}\mu_{F,l}\mu_{F,j-l}\sum_{i\geqslant n+1}c_i^{1-\rho}$$
$$\leqslant \mu_{F\star F,j}\sum_{i\geqslant n+1}c_i^{1-\rho}\,.$$

For t at least t_1, Lemma 5.2.3 implies

$$|\mathrm{D}^{j+k}\overline{F}_i(t)| \leqslant Mc_i^{\alpha-\epsilon}t^{-k-j}\overline{F}(t)\,.$$

So, the first double sum in (6.7.1) is at most

$$Mt^{-k}\overline{F}(t)\sum_{i\geqslant n+1}c_i^{1-\rho}\,.$$

To bound the second double sum, we use Lemma 6.3.5 to obtain, for j at least 1,

$$\mu_{G\setminus i,j} \leqslant \mu_{G,j} \leqslant C_{1-\rho}\mu_{F,j}\,.$$

Hence, the second double sum in (6.7.1) is at most

$$M\sum_{i\geqslant n+1}c_i^{\alpha-\epsilon}t^{-k}\overline{F}(t)\,.$$

This concludes the proof. ∎

The next lemma is the only place where the boundedness and continuity of $F^{(k)}$ is used. Its proof is an adaptation of that of Proposition 9.1.6 in Dudley (1989).

LEMMA 6.7.2. *Assume that F is k-times continuously differentiable on $(0,\infty)$, and that $F^{(k)}$ is bounded and in $L^1(\,\mathrm{d}x)$. Then, $\lim_{n\to\infty}G_n^{(k)} = G^{(k)}$ pointwise.*

PROOF. Write

$$G_n(t) = \int_0^t F_1(t-y)\,\mathrm{d}G_{n\setminus 1}(y)\,.$$

Since $F^{(k)}$ exists and is in $L^1(\,\mathrm{d}x)$, so is F_1. Then

$$G_n^{(k)}(t) = \int_0^t F_1^{(k)}(t-y)\,\mathrm{d}G_{n\setminus 1}(y)\,.$$

But the sequence $(G_{n\setminus 1})_{n\geqslant 1}$ converges weakly* to the continuous distribution function $G_{\setminus 1}$ and since $F_1^{(k)}$ is continuous and bounded, $G_n^{(k)}(t)$ converges to

$$\int_0^t F_1^{(k)}(t-y)\,\mathrm{d}G_{\setminus 1}(y) = \bigl(F_1 \star (G_{\setminus 1})\bigr)^{(k)}(t) = G^{(k)}(t)\,.$$
∎

To obtain Theorem 2.5.1 when the c_i's are nonnegative, it remains to do some rewriting of Lemma 6.6.5. For this purpose, define the sets of nonnegative sequences

$$\mathcal{C}_{\alpha,\omega,\gamma} = \{\, c \in [0,\infty)^{\mathbb{N}^*} : N_{\alpha,\gamma,\omega}(c) \leqslant 1 \,\}.$$

The following is exactly Theorem 2.5.1 but for the positive case and everywhere smooth distribution function.

PROPOSITION 6.7.3. *Let $\overline{F}$ be in $SR_{-\alpha,\omega}$ and let k, m be two nonnegative integers with $k+m$ smaller than ω and m smaller than α. Let γ be a positive real number smaller than both 1 and $\omega - k - m$. Assume that $F^{(k)}$ is continuous and bounded. There exists a function $\eta(\cdot)$ which tends to 0 at infinity and a real number t_{13} such that for any t at least t_{13} and any sequence c in $\mathcal{C}_{\alpha,\omega,\gamma}$*

$$|\overline{G}^{(k)} - \mathcal{A}_m \overline{G}^{(k)}|_{k+m,t} \leqslant \eta(t).$$

PROOF. For ϵ positive, define $\rho = (1/2) \wedge (\alpha/(\alpha+\omega))$. Let c be a sequence in $\mathcal{C}_{\alpha,\omega,\gamma}$. By definition of $\mathcal{C}_{\alpha,\omega,\gamma}$ the series $C_{\rho\gamma}$ is at most 1. Moreover, by taking ϵ small enough, $C_{\alpha-\epsilon}$ is at most $C_{\rho\gamma}$, hence at most 1.

Since $k+m+\gamma$ is less than ω, the constant ρ is less than $\alpha/(\alpha+\gamma+k+m)$, implying that $\alpha - \rho(\alpha+k+m)$ is less than $\rho\gamma$. Hence, for ϵ small enough $C_{\alpha-(\alpha+k+m)\rho-\epsilon}$ is at most 1.

To conclude the proof, Lemmas 6.7.1 and 6.7.2 show that

$$\overline{G}^{(k)} - \mathcal{A}_m \overline{G}^{(k)} = \lim_{n \to \infty} \overline{G}_n^{(k)} - \mathcal{A}_m \overline{G}_n^{(k)}$$

pointwise. But for any n at least 2 and any t at least t_1, Lemma 6.6.5 yields

$$|\overline{G}_n^{(k)}(t) - \mathcal{A}_m \overline{G}_n^{(k)}(t)| \leqslant \eta(t) t^{-m-k} \overline{F}(t).$$

Hence the same inequality holds with $\overline{G}^{(k)}$. ∎

CHAPTER 7

Removing the sign restriction on the random variables.

In section 6, we assumed that the random variables are positive. Our goal in this section is to remove this assumption. But we will keep the assumption that $F^{(k)}$ and $M_{-1}F^{(k)}$ exist, are bounded and continuous on the whole positive half-line. The basic argument consists in conditioning. To be specific, define the random set of all indices corresponding to a nonpositive random variable, that is

$$I = \{\, i \in \mathbb{N}^* : X_i \leqslant 0 \,\}.$$

Write H_I for the distribution function of $\sum_{i \in I} c_i X_i$ and G_I for that of $\sum_{i \in \mathbb{N}^* \setminus I} c_i X_i$. If i is in I, we write $H_{I \setminus i}$ for the distribution function of $\sum_{j \in I \setminus \{i\}} c_j X_j$. If i is not in I, we write $G_{I \setminus i}$ for the distribution function of $\sum_{j \in (\mathbb{N}^* \setminus I) \setminus \{i\}} c_j X_j$.

Let F_+ (respectively F_-) be the distribution function of X_1 say, given that X_1 is positive (respectively nonpositive), that is

$$\overline{F_+} = \overline{F}/\overline{F}(0) \quad \text{on} \quad (0, \infty)$$

and

$$F_- = F/F(0) \quad \text{on} \quad (-\infty, 0].$$

We write $F_{+,i}$ (respectively $F_{-,i}$) for $M_{c_i}F_+$ (respectively $M_{c_i}F_-$). Note that there is no ambiguity in this notation for when c is nonnegative, $(M_cF)_+ = M_c(F_+)$ and $(M_cF)_- = M_c(F_-)$. Given I, the results of section 6 can be applied to both $M_{-1}H_I$ and G_I, using respectively the distribution functions $M_{-1}F_-$ and F_+ instead of F. The distribution function of the whole series $\langle c, X \rangle$ is $EH_I \star G_I$. The identity

$$\overline{H_I \star G_I}(t) = \int_{-\infty}^{0} \overline{G_I}(t - x) \, \mathrm{d}H_I(x),$$

suggests the relevance to the operator

$$U_H h(t) = \int_{-\infty}^{0} h(t - x) \, \mathrm{d}H(x),$$

in terms of which $\overline{H_I \star G_I} = U_{H_I}\overline{G_I}$. We then need to prove the analogue of some of the results of section 6 on $T_{F,\eta}$, but now for the operator U_H. The first one which we will prove implies $\mathrm{D}^k U_H \overline{G} = U_H \overline{G}^{(k)}$. It is then clear how the proof of the asymptotic expansion for the distribution function K of $\langle c, X \rangle$ and its derivatives will go. Indeed, we will have

$$\overline{K}^{(k)} = EU_{H_I}\overline{G}_I^{(k)} = EU_{H_I}\mathcal{A}_m\overline{G}_I^{(k)} + EU_{H_I}(\overline{G}_I^{(k)} - \mathcal{A}_m\overline{G}_I^{(k)}). \qquad (7.0.1)$$

We will see that U_H is a contraction for the right norm. Hence, equality (7.0.1) combined with the results of section 6 gives the asymptotic expansion

$$\overline{K}^{(k)} \sim EU_{H_I}\mathcal{A}_m\overline{G}_I^{(k)}.$$

We will be able to approximate U_{H_I} by L_{H_I}. Because the error term is bounded uniformly with respect to the sequence $(c_i)_{i\geqslant 1}$, we will be able to permute expectation and asymptotic expansions.

7.1. Elementary properties of U_H.

Since we are interested in expansions of derivatives, our first elementary result deals with the composition of the derivative and U_H. It asserts that those two operators commute whenever acting on sufficiently regular functions.

LEMMA 7.1.1. *Let h be a function on the nonnegative half-line with Lebesgue integrable k-th derivative, and such that for any nonnegative l at most k,*

$$\lim_{t\to\infty} h^{(l)}(t) = 0.$$

Then, $D^k U_H h = U_H D^k h$ almost everywhere.

PROOF. By induction it suffices to prove the result for k equal 1. We write

$$U_H h(t) = -\int\int h'(y-x)\mathbb{1}\{y > t\}\,dy\,dH(x).$$

The function $U_H h'$ is Lebesgue integrable, for

$$\int \left|\int h'(y-x)\,dH(x)\right|dy \leqslant \int\int |h'(y-x)|\,dy\,dH(x) = |h'|_{L^1(dx)}.$$

Applying Fubini's theorem,

$$U_H h(t) = -\int_t^\infty \int_{-\infty}^0 h'(y-x)\,dH(x)\,dy = -\int_t^\infty U_H h'(y)\,dy.$$

Since we proved that $U_H h'$ is Lebesgue integrable, it is almost everywhere the derivative of $U_H h$. ∎

In particular, if $F^{(k)}$ is continuous and bounded and i is not in I, then the analogue to the expression obtained in the proof of Lemma 6.7.2 is

$$\overline{G}_I^{(k)} = \int_0^t \overline{F}_{+,i}^{(k)}(t-y)\,dG_{I\setminus i}(y). \qquad (7.1.1)$$

Thus, $\overline{G}_I^{(k)}$ is bounded and integrable with respect to the Lebesgue measure. Moreover, since $\overline{F}$ is smoothly varying with negative exponent $-\alpha$, its k-th derivative tends to 0 at infinity. Then (7.1.1) shows that $\overline{G}_I^{(l)}$ tends to 0 at infinity for any nonnegative l at most k. Consequently,

$$\overline{H_I \star G_I}^{(k)} = U_{H_I} \overline{G}_I^{(k)}.$$

Our next result allows us to replace $\overline{G}_I^{(k)}$ by its asymptotic expansion when looking for an expansion for $U_{H_I} \overline{G}_I^{(k)}$.

LEMMA 7.1.2. *For any nonnegative t, the operator U_H is a contraction with respect to the norms $|\cdot|_{p,t}$.*

PROOF. Let h be a function whose $|\cdot|_{p,t}$-norm is finite. Since $t \mapsto t^p/\overline{F}(t)$ is nondecreasing on the nonnegative half-line

$$\left| \frac{t^p}{\overline{F}(t)} U_H h(t) \right| \leq \int_{-\infty}^0 \frac{t^p}{\overline{F}(t)} |h(t-x)| \, dH(x)$$

$$\leq \int_{-\infty}^0 \frac{(t-x)^p}{\overline{F}(t-x)} |h(t-x)| \, dH(x).$$

The integrand is at most $|h|_{p,t}$; so is the integral, for H is a distribution function. ∎

7.2. Basic expansion of U_H.

We now show that U_H has an expansion $L_{H,m}$, and therefore behaves similarly to $T_{F,\eta}$. Indeed, the next lemma should be compared to Theorem 6.4.1.

LEMMA 7.2.1. *Let m be a positive integer and let r be in $[0,1)$. Let H be a distribution function on the nonpositive half-line with finite m-th moment. If h is smoothly varying of order $m+r$, then*

$$|(U_H - L_{H,m})h|(t) \leq \sum_{0 \leq j \leq m} \frac{|h^{(j)}(t)|}{j!} \int_{-\infty}^{-t/2} |x|^j \, dH(x)$$

$$+ \frac{|h^{(m)}(t)|}{t^r m!} \int_{-t/2}^0 \overline{\Delta}^r_{t, |x|/t}(h^{(m)}) |x|^{m+r} \, dH(x)$$

$$+ H(-t/2) \sup_{s \geq t} |h(s)|.$$

PROOF. We first have the inequality

$$\left| U_H h(t) - \int_{-t/2}^0 h(t-x) \, dH(x) \right| \leq \int_{-\infty}^{-t/2} |h(t-x)| \, dH(x)$$

$$\leq H(-t/2) \sup_{s \geq t} |h(s)|.$$

Applying Proposition 5.2.5, we have

$$\left| \int_{-t/2}^{0} \left(h(t-x) - \sum_{0 \leqslant j \leqslant m} \frac{(-1)^j}{j!} x^j h^{(j)}(t) \right) dH(x) \right|$$

$$\leqslant \int_{-t/2}^{0} \frac{|x|^{m+r}}{t^r} \frac{|h^{(m)}(t)|}{m!} \overline{\Delta}_{t,|x|/t}^r (h^{(m)}) \, dH(x).$$

The result follows. ∎

7.3. A technical lemma.

Looking at equality (7.0.1), we need to approximate $U_{H_I} \mathcal{A}_m \overline{G}_I^{(k)}$. The following result, an analogue to Lemma 6.5.4, will do. Since $M_{-1} H_I$ has the same properties as the distribution function G studied in section 6, it satisfies the assumptions of the next lemma. Note that even if $M_{-1} F$ is not regularly varying, the assumption $\overline{M_{-1}F} = O(\overline{F})$ and Lemma 6.3.1— for its part which does not require regular variation — ensure that H_I still satisfies the assumption of the following Lemma.

LEMMA 7.3.1. *Assume that there exist constants M_0 and t_{14} such that for any integer k less than α and any t at least t_{14},*

$$\int_{-\infty}^{-t} |x|^k \, dH(x) \leqslant M_0 t^k \overline{F}(-t)$$

and for any nonnegative integer j at most m, the moments $\mu_{M_{-1}H, j+r}$ are at most M_0. Then, there exist a positive M and a t_{15} such that for any $G = \star_{i \geqslant 1} F_i$ as in section 6 (that is the distribution function of an infinite weighted sum of nonnegative random variables with nonnegative weights with the assumption of section 6 satisfied), for any t at least t_{15},

$$|U_H \mathcal{A}_m \overline{G}^{(k)} - \sum_{i \geqslant 1} L_{H \star G \setminus i, m} \overline{F}_i^{(k)}|_{k+m, t} \leqslant M M_0 C_{\alpha - \epsilon} \Big(\overline{\Delta}_{t, v(t)/t}^r (\overline{F}^{(k+m)})$$

$$+ \overline{\Delta}_{t, 1/2}^r (\overline{F}^{(k+m)}) v(t)^{m+r} \overline{F}(-v(t)) + t^m H(-t/2) \Big).$$

The proof uses an auxiliary result which we state as a claim, very much as we did in the proof of Lemma 6.5.4.

CLAIM. *For any nonnegative t, any distribution function H supported on the nonpositive half-line and K supported on the nonnegative half-line,*

$$|U_H L_K h - L_{H \star K} h|(t) \leqslant \sum_{0 \leqslant s \leqslant m} \frac{|h^{(s)}(t)|}{s!} \sum_{0 \leqslant j \leqslant s} \binom{s}{j} \mu_{K,j} \int_{-\infty}^{-t/2} |x|^{s-j} \, dH(x)$$

$$+ \frac{|h^{(m)}(t)|}{t^r m!} \sum_{0 \leqslant j \leqslant m} \binom{m}{j} \mu_{K,j} \int_{-t/2}^{0} \overline{\Delta}_{t,|x|/t}^r (h^{(m)}) |x|^{m-j+r} \, dH(x)$$

$$+ H(-t/2) \sum_{0 \leqslant j \leqslant m} \frac{\mu_{K,j}}{j!} \sup_{s \geqslant t} |h^{(j)}(s)|.$$

PROOF. By linearity of $\mathcal{U}_H$,
$$\mathcal{U}_H L_K = \sum_{0 \leqslant j \leqslant m} \frac{(-1)^j}{j!} \mu_{K,j} \mathcal{U}_H \mathrm{D}^j \,.$$

Applying Lemma 7.2.1 to bound $(\mathcal{U}_H - L_{H,m-j})\mathrm{D}^j h$, we obtain

$$\left|\mathcal{U}_H L_K h - \sum_{0 \leqslant j \leqslant m} \frac{(-1)^j}{j!} \mu_{K,j} L_{H,m-j} h^{(j)}\right|(t)$$

$$\leqslant \sum_{0 \leqslant j \leqslant m} \frac{\mu_{K,j}}{j!} \sum_{0 \leqslant l \leqslant m-j} \frac{|h^{(l+j)}(t)|}{l!} \int_{-\infty}^{-t/2} |x|^l \,\mathrm{d}H(x)$$

$$+ \sum_{0 \leqslant j \leqslant m} \frac{\mu_{K,j}}{j!} \frac{|h^{(m)}(t)|}{t^r(m-j)!} \int_{-t/2}^{0} \overline{\Delta}_{t,|x|/t}^r(h^{(m)})|x|^{m-j+r} \,\mathrm{d}H(x)$$

$$+ \sum_{0 \leqslant j \leqslant m} \frac{\mu_{K,j}}{j!} H(-t/2) \sup_{s \geqslant t} |h^{(j)}(s)| \,.$$

Again, Lemma 5.1.3 shows that the left hand side of this inequality is the absolute value of $(\mathcal{U}_H L_K - L_{H \star K})h$ evaluated at t. Setting $s = l + j$, the right hand side is exactly the upper bound given in the claim. ∎

PROOF (of Lemma 7.3.1). The triangle inequality implies the pointwise inequality

$$\left|\mathcal{U}_H \mathcal{A}_m \overline{G}^{(k)} - \sum_{i \geqslant 1} L_{H \star G_{\setminus i},m} \overline{F}_i^{(k)}\right| \leqslant \sum_{i \geqslant 1} |\mathcal{U}_H L_{G_{\setminus i},m} \overline{F}_i^{(k)} - L_{H \star G_{\setminus i},m} \overline{F}_i^{(k)}| \,. \quad (7.3.1)$$

The claim yields, for any positive t,

$$|\mathcal{U}_H L_{G_{\setminus i}} \overline{F}_i^{(k)} - L_{H \star G_{\setminus i}} \overline{F}_i^{(k)}|(t)$$

$$\leqslant \sum_{0 \leqslant s \leqslant m} \frac{|\overline{F}_i^{(k+s)}(t)|}{s!} \sum_{0 \leqslant j \leqslant s} \binom{s}{j} \mu_{G_{\setminus i},j} \int_{-\infty}^{-t/2} |x|^{s-j} \,\mathrm{d}H(x)$$

$$+ \frac{|\overline{F}_i^{(k+m)}(t)|}{t^r m!} \sum_{0 \leqslant j \leqslant m} \binom{m}{j} \mu_{G_{\setminus i},j} \int_{-t/2}^{0} \overline{\Delta}_{t,|x|/t}^r(\overline{F}_i^{(k+m)})|x|^{m-j+r} \,\mathrm{d}H(x)$$

$$+ H(-t/2) \sum_{0 \leqslant j \leqslant m} \frac{\mu_{G_{\setminus i},j}}{j!} \sup_{s \geqslant t} |\overline{F}_i^{(k+j)}(s)| \,. \quad (7.3.2)$$

We use the same estimates as in the proof of Lemma 6.5.4. So, Lemma 5.2.3 yields

$$|\overline{F}_i^{(k+s)}(t)| \leqslant M c_i^{\alpha - \epsilon} t^{-k-s} \overline{F}(t) \,,$$

and we also have the moment bound

$$\mu_{G_{\setminus i},j} \leqslant \mu_{G,j} \leqslant M \,.$$

In (7.3.2), the term
$$\int_{-t/2}^{0} \overline{\Delta}^r_{t,|x|/t}(\overline{F}_i^{(k+m)})|x|^{m-j+r}\,\mathrm{d}H(x)$$
is rewritten as a sum of an integral over $[-t/2, -v(t))$ plus an integral over $[-v(t), 0]$. In this decomposition, the second integral is at most
$$\overline{\Delta}^r_{t,v(t)/t}(\overline{F}^{(k+m)})\mu_{M_{-1}H,m-j+r}\,,$$
while the first one is at most
$$\overline{\Delta}^r_{t,1/2}(\overline{F}^{(k+m)}) \int_{-\infty}^{-v(t)} |x|^{m-j+r}\,\mathrm{d}H(x)\,.$$

By assumption, for t at least $2t_{14}$, the inequality
$$\int_{-\infty}^{-t/2} |x|^{s-j}\,\mathrm{d}H(x) \leqslant M_0 (t/2)^{s-j} F(-t/2)$$
holds true. It follows that for t more than some t'_{14}, the right hand side of (7.3.2) is at most
$$Mc_i^{\alpha-\epsilon}t^{-k}\overline{F}(t)F(-t/2) + Mc_i^{\alpha-\epsilon}t^{-k-m-r}\overline{F}(t)\Big(\overline{\Delta}^r_{t,v(t)/t}(\overline{F}^{(k+m)})$$
$$+ \overline{\Delta}^r_{t,1/2}(\overline{F}^{(k+m)})v(t)^{m+r}F\big(-v(t)\big)\Big)$$
$$+ MH(-t/2)c_i^{\alpha-\epsilon}t^{-k}\overline{F}(t)\,.$$

Consequently, for t at least $2t_{14} \wedge t'_{14}$, (7.3.1) is at most
$$MC_{\alpha-\epsilon}t^{-k-m}\overline{F}(t)\Big(t^m F(-t/2) + \overline{\Delta}^r_{t,v(t)/t}(\overline{F}^{(k+m)})$$
$$+ \overline{\Delta}^r_{t,1/2}(\overline{F}^{(k+m)})v(t)^{m+r}F\big(-v(t)\big) + t^m H(-t/2)\Big)\,.$$

For t large enough, $t^m F(-t/2)$ is at most $v(t)^{m+r} F\big(-v(t)\big)$. This concludes the proof of the lemma. ∎

7.4. Conditional expansion and removing conditioning.

For any I, the distribution functions $M_{-1}H_I$ and G_I have asymptotic expansions given by the results of section 6. In particular, H_I satisfies the assumptions of Lemma 7.3.1 with a constant M_0 independent of both I and the sequence $c = (c_i)_{i \geqslant 1}$ for which $N_{\alpha,\omega,\gamma}(c)$ is at most 1.

The result of section 6 shows that for some function $\eta(\cdot)$ which tends to 0 at infinity, for any t at least some t_{13}, for any set I and any sequence c with $N_{\alpha,\omega,\gamma}(c)$ at most 1,
$$|\overline{G}_I^{(k)} - \mathcal{A}_m \overline{G}_I^{(k)}|_{k+m,t} \leqslant \eta(t)\,,$$

where
$$\mathcal{A}_m \overline{G}_I^{(k)} = \sum_{i \in \mathbb{N}^* \setminus I} L_{G_{I \setminus i}} \overline{F}_{+,i}^{(k)}.$$

Next, Lemma 7.1.1 shows that $\overline{H_I \star G_I}^{(k)} = U_{H_I} \overline{G}_I^{(k)}$, which justifies equality (7.0.1). Then, Lemma 7.1.2 and the result of section 6 yield

$$|\overline{K}^{(k)} - E U_{H_I} \mathcal{A}_m \overline{G}_I^{(k)}|_{k+m,t} \leqslant \eta(t).$$

Using Lemma 7.3.1, we conclude that there exists a function $\eta^*(\cdot)$ with limit 0 at infinity such that

$$\left| \overline{K}^{(k)} - E \sum_{i \in \mathbb{N}^* \setminus I} L_{H_I \star G_{I \setminus i}} \overline{F}_{+,i}^{(k)} \right|_{k+m,t} \leqslant \eta^*(t).$$

To calculate the expectation involved in the inequality, we rewrite it as

$$\sum_{i \geqslant 1} E\big(\mathbb{1}\{X_i > 0\} L_{H_I \star G_{I \setminus i}} \big) \overline{F}_{+,i},$$

or, equivalently, since the X_i's are independent and identically distributed, as

$$\sum_{i \geqslant 1} E(L_{H_I \star G_{I \setminus i}} \mid X_i > 0) \overline{F}(0) \overline{F}_{+,i}^{(k)}.$$

Conditionally upon X_i being positive, $H_I \star G_{I \setminus i}$ is the distribution function of $\sum_{\substack{j \geqslant 1 \\ j \neq i}} c_j X_j$ given I. Thus, the X_i's being independent,

$$E(L_{H_I \star G_{I \setminus i}} \mid X_i > 0) = L_{G_{\setminus i}}.$$

Next, $\overline{F}(0) \overline{F}_{+,i}^{(k)} = \overline{F}_i^{(k)}$ on the positive half-line, and so the expectation that we wanted to calculate is simply $\sum_{i \geqslant 1} L_{G_{\setminus i}} \overline{F}_i^{(k)}$. This proves Theorem 2.5.1 when the X_i's may assume arbitrary sign and the constants c_i's are nonnegative and F is smooth.

CHAPTER 8

Removing the sign restriction on the constants.

So far, we proved Theorem 2.5.1 assuming that the c_i's are nonnegative. To drop this requirement is now rather easy, but unfortunately requires some checking which is very much in the flavor of either section 6 or 7. So we will give the details only for part of the proof. In this section, we keep assuming that $\overline{F}^{(k)}$ and $\overline{M_{-1}F}^{(k)}$ exist, are bounded and continuous on the positive half-line.

We define the set J of all indices pertaining to a positive constant c_i, that is

$$J = \{\, i \in \mathbb{N}^* : c_i > 0 \,\}.$$

Define G (respectively H) to be the distribution function of the series $\sum_{i \in J} c_i X_i$ (respectively $\sum_{i \in \mathbb{N}^* \setminus J} c_i X_i$). When i belongs to J, we also define $G_{\setminus i}$ to be the distribution function of $\sum_{j \in J \setminus \{i\}} c_j X_j$ and when i does not belong to J, we set $H_{\setminus i}$ to be the distribution function of $\sum_{j \in (\mathbb{N}^* \setminus J) \setminus \{i\}} c_j X_j$.

The tail expansion of $\overline{G}^{(k)}$ is given by the result of section 7. That of $\overline{H}^{(k)}$ follows in the same way, since

$$\sum_{i \in \mathbb{N}^* \setminus J} c_i X_i = \sum_{i \in \mathbb{N}^* \setminus J} (-c_i)(-X_i).$$

Now, the distribution function K of $\langle c, X \rangle$ is the convolution $G \star H$.

To obtain an asymptotic expansion of $\overline{G \star H}^{(k)}$, we have two strategies. One is to decompose the distribution functions by conditioning with respect to signs, as we did in section 7; alternatively, we could proceed along the lines of section 6. Both routes are about equal length, and we go for the latter.

Our starting point is Proposition 6.1.1 which asserts that

$$\overline{K} = T_{G,1/2}\overline{H} + T_{H,1/2}\overline{G} + M_2 \overline{G} M_2 \overline{H}$$

and for a positive integer k,

$$\overline{K}^{(k)} = T_{G,1/2}\overline{H}^{(k)} + T_{H,1/2}\overline{G}^{(k)} + \sum_{1 \leqslant i \leqslant k-1} M_2 \overline{G}^{(i)} M_2 \overline{H}^{(k-i)}. \qquad (8.0.1)$$

8.1. Neglecting terms involving the multiplication operators.

Starting with (8.0.1), the purpose of this subsection is to show that an expansion of $\overline{K}^{(k)}$ can be obtained from the ones for $T_{G,1/2}\overline{H}^{(k)}$ and $T_{H,1/2}\overline{G}^{(k)}$, or, equivalently, that the terms $M_2 \overline{G}^{(i)} M_2 \overline{H}^{(k-i)}$ can be neglected.

Our first lemma is an analogue of Lemma 7.1.2. The operators M_c are not contractions, but for our problems, they behave very much as if they were bounded, which is good enough.

LEMMA 8.1.1. *There exists t_{16} such that for any positive real number c and any positive t with $(t/c) \wedge t$ at least t_{16},*

$$|M_c h|_{p,t} \leqslant (c^{p+\alpha+1} \vee c^{p+\alpha-1}) |h|_{p,t/2}.$$

PROOF. We simply write

$$\frac{s^p}{\overline{F}(s)} h(s/c) = \frac{(s/c)^p h(s/c)}{\overline{F}(s/c)} c^p \frac{\overline{F}(s/c)}{\overline{F}(s)},$$

from which we deduce

$$|M_c h|_{p,t} \leqslant c^p |h|_{p,t/c} \sup_{s \geqslant t} \overline{F}(s/c)/\overline{F}(s).$$

We conclude by using Lemma 5.2.4. ∎

Our next lemma states a rather obvious property of the norm $|\cdot|_{p,t}$.

LEMMA 8.1.2. *For any positive reals p, q and t,*

$$|fg|_{p+q+m,t} \leqslant |f|_{p,t} |g|_{q,t} \sup_{s \geqslant t} s^m \overline{F}(s).$$

PROOF. The result follows from the identity

$$\frac{s^{p+q+m}}{\overline{F}(s)} f(s) g(s) = \frac{s^p}{\overline{F}(s)} f(s) \frac{s^q}{\overline{F}(s)} g(s) s^m \overline{F}(s). \qquad \blacksquare$$

Our first two lemmas in this subsection imply the following bound.

LEMMA 8.1.3. *Assume that the hypotheses of Theorem 2.5.1 hold. Suppose also that $F^{(k)}$ exists, is continuous and bounded. Then, there exists a function $\eta^*(\cdot)$ with limit 0 at infinity and a positive number t_{17} such that for any t at least t_{17} and any sequence c with $N_{\alpha,\omega,\gamma}(c)$ at most 1,*

$$\left| \sum_{1 \leqslant i \leqslant k-1} M_2 \overline{G}^{(i)} M_2 \overline{H}^{(k-i)} \right|_{k+m,t} \leqslant \eta^*(t).$$

PROOF. The triangle inequality, Lemmas 8.1.1 and 8.1.2 yield for t at least 1,

$$\left| \sum_{1 \leqslant i \leqslant k-1} M_2 \overline{G}^{(i)} M_2 \overline{H}^{(k-i)} \right|_{k+m,t}$$

$$\leqslant \sum_{1 \leqslant i \leqslant k-1} 2^{k+m+\alpha+1} |\overline{G}^{(i)}|_{i,t/2} |\overline{H}^{(k-i)}|_{k-i,t/2} \sup_{s \geqslant t/2} s^m \overline{F}(s).$$

From the result of section 7,

$$|\overline{G}^{(i)} - \mathcal{A}_0\overline{G}^{(i)}|_{i,t/2} \leqslant \eta(t/2).$$

Since

$$\mathcal{A}_0\overline{G}^{(i)} = \sum_{n\in J}\overline{F}_n^{(i)},$$

the triangle inequality and lemma 5.2.3 imply that for t at least t_1,

$$|\mathcal{A}_0\overline{G}^{(i)}(t)| \leqslant M\sum_{n\in J} c_n^{\alpha-\epsilon} t^{-i}\overline{F}(t).$$

Consequently,

$$|\overline{G}^{(i)}|_{i,t/2} \leqslant \eta(t/2) + MC_{\alpha-\epsilon}.$$

We have a similar bound for $|H^{(k-i)}|_{k+m,t/2}$. Since $s^m\overline{F}(s) = o(1)$ as s tends to infinity, the result follows. ∎

8.2. Substituting $\overline{H}^{(k)}$ and $\overline{G}^{(k)}$ by their expansions.

Looking at equality (8.0.1), we would like to use Theorem 6.4.1 to conclude that $T_{G,1/2}\overline{H}^{(k)}$ has asymptotic expansion $L_{G_m}\overline{H}^{(k)}$. This is not possible since we do not know if $\overline{H}^{(k)}$ is smoothly varying — the problem is not so much about the regular variation part of the smoothly varying condition, but more about the continuity of $\overline{H}^{(k+m)}$. The trick is to replace $\overline{H}^{(k)}$ first by its asymptotic expansion. Since the expansion involves explicitly $\overline{F}$ and its derivatives, we will be able to study the action of $T_{G,1/2}$ on the expansion. The first step is to show that $T_{G,1/2}$ behaves like a bounded operator in our problem (though it is not.)

LEMMA 8.2.1. *For t at least t_1,*

$$|T_{G,1/2}h|_{p,t} \leqslant 2^{p+\alpha+1}|h|_{p,t/2}.$$

PROOF. The proof follows closely that of Lemma 6.5.2. ∎

This lemma and the result of section 7 imply

$$|T_{G,1/2}\overline{H}^{(k)} - T_{G,1/2}\mathcal{A}_m\overline{H}^{(k)}|_{k+m,t} \leqslant 2^{k+m+\alpha+1}\eta(t/2).$$

Of course a similar inequality holds if we permute G and H.

To conclude the proof of Theorem 2.5.1 when F is smooth, we write

$$T_{G,1/2}\mathcal{A}_m\overline{H}^{(k)} = \sum_{n\in\mathbb{N}^*\setminus J} T_{G,1/2}L_{H_{\setminus n},m}\overline{F}_n^{(k)}.$$

Since $L_{H\backslash n, m}\overline{F}_n^{(k)}$ involves derivatives of $\overline{F}_n$, it then suffices to study the expansion of $T_{G,1/2}\overline{F}_n^{(k)}$ and have a good error bound in this expansion. This is now straightfoward. Indeed, write

$$T_{G,1/2}\overline{F}_n^{(k)} = \int_{-\infty}^{-t/2} \overline{F}_n(t-x)\,dG(x) + \int_{-t/2}^{t/2} \overline{F}_n^{(k)}(t-x)\,dG(x).$$

We then bound

$$\left| \int_{-\infty}^{-t/2} \overline{F}_n^{(k)}(t-x)\,dG(x) \right| \leqslant G(-t/2) \sup_{s \geqslant t/2} |\overline{F}_n^{(k)}(s)|$$
$$\leqslant M|c_n|^k G(-t/2) t^{-k} F(-t/2),$$

and we expand

$$\int_{-t/2}^{t/2} \overline{F}_n^{(k)}(t-x)\,dG(x)$$

into $L_{G,m}\overline{F}^{(k)}$ as we did in sections 6 and 7.

This completes the proof of Theorem 2.5.1 when $F^{(k)}$ is continuous and bounded.

CHAPTER 9

Removing the smoothness assumption.

We now want to remove the assumption that $\overline{F}^{(k)}$ exist, is continuous and bounded on the whole real line.

In Theorem 2.5.1, the membership of $\overline{F}$ and $\overline{M_{-1}F}$ to $SR_{-\alpha,\omega}$ ensures that there exists some A more than 2 say such that $\overline{F}^{(k)}, \overline{M_{-1}F}^{(k)}$ exist, are continuous and bounded on (A, ∞). Thus, we can write $F = (1-p)F_0 + pF_1$ where F_0 is concentrated on $(-A-1, A+1)$ and F_1 is concentrated on $(-A, A)^c$, and such that $F_1^{(k)}$ and $\overline{M_{-1}F_1}^{(k)}$ exist, are continuous and bounded on the whole real line. Let ϵ_i, $i \in \mathbb{N}$, be a sequence of independent Bernoulli random variables with mean p. Let I be the random set of integers i for which ϵ_i is 1. Given I, let G_I be $\star_{i \in \mathbb{Z} \setminus I} M_{c_i} F$ and H_I be $\star_{i \in I} M_{c_i} F_0$. Then $\star_{i \in \mathbb{Z}} M_{c_i} F_1$ is $EG_I \star H_I$. Since the c_i's are summable, the support of the distributions pertaining to G_I is included in $|c|_1[-A-1, A+1]$ no matter what I is. It follows that on $2|c|_1(A+1, \infty)$,

$$\overline{G_I \star H_I} = T_{G_I}\overline{H_I}.$$

The expansion for $\overline{G_I \star H_I}$ follows from that of H_I using the same arguments that we used in sections 7 and 8. This concludes the proof of Theorem 2.5.1.

APPENDIX

Maple code

The goal in writing the following Maple (version 9) code was to see to what extent the tail calculus described in section 3.3 could be automated. With this code, the user enters a distribution function having an expansion in some power of t^{-1} and the integer m occurring in Theorem 2.5.1. The program then generates the coefficients of the asymptotic expansion of F_C (with the notation of Theorem 2.5.1), assuming that the c_i's are positive. Throughout this appendix, we write G for F_C.

The example which we ran here is that of the Burr distribution in subsection 3.4.

The tail $\overline{F}$ is oF (for 'overlined F'). One needs to specify the parameter α, which is named palpha and m.

```
restart; with(LinearAlgebra):
pgamma:=10: ptau:=3/2:
oF:=(1+(t^(ptau))/beta)^(-pgamma);
palpha:=ptau*pgamma; m:=4;
```

We then expand the tail of F. We write the tail expansion as a polynomial of $x = 1/t$.

```
equiF:=subs(t=1/x,convert(asympt( oF,t,palpha+m),polynom)):
```

We build the set of powers of x involved in this expansion.

```
index_set:={}:
for i from 1 by 1 to nops(equiF) do
  a:=op(i,equiF):
  index_set:=index_set union {limit(log(a)/log(x),x=infinity)}:
end do:
power_list:=sort(convert(index_set,'list')):
```

We complete this list by the augmentation procedure described in section 3.2.

```
for i in op(index_set) do
  for j from i by 1 while  j <= palpha+m do
    index_set:=index_set union {j};
  end do;
end do;
index_list:=sort(convert(index_set,'list')):
```

Then we calculate the dimension of the vector space we will work with.

```
vdim:=nops(index_list):
```

We calculate the vector $p_{\overline{F}}$. Unfortunately, with `Maple`, it is easy to work with polynomials, but it has no useful command to work with monomials of noninteger degree. That induces the following code where we obtained the monomials one by one.

```
pF:=Matrix(vdim,1,readonly=false):
equiF_tmp:=equiF:
for i from 1 by 1 to nops(power_list) do
  a:=op(i,equiF_tmp):
  xpower:=limit(log(a)/log(x),x=infinity):
  member(xpower,index_list,'k'):
  pF[k,1]:=a/x^xpower:
end do:
```

The next step is to build the matrices $\mathcal{D}$ and $\mathcal{M}_c$, which we call `Dmat` and `Mcmat` in the code.

```
Dmat:=Matrix(vdim,readonly=false):
for i from 1 by 1 to vdim-1 do
  for j from i+1 by 1 to vdim do
    if evalb(index_list[j]-index_list[i]=1)
      then Dmat[j,i]:=-index_list[i]; end if;
end do;end do;
Mcmat:=Matrix(vdim,readonly=false):
for i from 1 by 1 to vdim do
  Mcmat[i,i]:=c^index_list[i]:
end do:
```

We construct the Laplace character of F in its matrix form $\mathcal{L}_F$. The matrix is `LFmat`.

```
LFmat:=Matrix(vdim,readonly=false):
temp_mat:=Matrix(vdim,readonly=false):
for i from 1 by 1 to vdim do temp_mat[i,i]:=1: end do:
mu[0]:=1:
for k from 0 by 1 to m do
  LFmat:=(LFmat+((-1)^k*temp_mat*(c^k)*mu[k]/k!)):
  temp_mat:=(temp_mat.Dmat):
end do:
```

Its inverse is

```
LFmat_inv:=MatrixInverse(LFmat,method='subs'):
```

We write the expansion for G as $\mathcal{L}_G \sum_{i\in\mathbb{Z}} \mathcal{L}_F^{-1} p_{\overline{M_{c_i}F}}$. We calculate the generic summands $\mathcal{L}_F^{-1} p_{\overline{M_c F}}$.

```
Msum:=LFmat_inv.Mcmat.pF:
```

To sum these summands amounts to substitute c^p by C_p in this sum, which we do now.

```
for i from 1 by 1 to vdim do
  Msum[i,1]:=expand(collect(Msum[i,1],c),c):
end do:
for i from 1 by 1 to vdim do
  a:=Msum[i,1]:
  for j from 1 by 1 to nops(a) do
    if is ( c in indets(op(j,a)) )
    then
      p:=limit(log(op(j,a))/log(c),c=infinity):
      Msum[i,1]:=subs(c^p=C[p],Msum[i,1]):
    end if:
  end do:
end do:
```

To calculate the Laplace character of G, we first obtain its moments. The algorithm is that described in Barbe and McCormick (2005). These moments are coded as `muG[k]`.

```
for k from 0 by 1 to m do
  Q[k]:=t^k*c^k*mu[k]/k!:
end do:
P1:=add(Q[k],k=0..m):
P2:=convert(series(ln(P1),t=0,m+1),polynom):
P3:=add(C[k]*coeff(P2,c,k),k=0..m):
P4:=convert(series(exp(P3),t=0,m+1),polynom):
for k from 0 by 1 to m do:
  muG[k]:=k!*coeff(P4,t,k):
end do:
```

We construct the Laplace character of G.

```
LGmat:=Matrix(vdim,readonly=false):
temp_mat:=Matrix(vdim,readonly=false):
for i from 1 by 1 to vdim do temp_mat[i,i]:=1: end do:
for k from 0 by 1 to m do
  LGmat:=(LGmat+((-1)^k*temp_mat*muG[k]/k!)):
  temp_mat:=(temp_mat.Dmat):
end do:
```

And finally obtain the coefficient of the tail, that is $p_{\overline{G}}$.

```
tail:=LGmat.Msum:
```

In our example for the Burr distribution, these coefficients are messy. The remaining code makes them nicer looking; at least looking good enough so that they can be written as we did.

The first step in the simplification is to substitute the centered moments for the noncentered ones. In the code, we write `kappa[k]` for the k-the centered moment. We express it as a function of the noncentered ones.

```
for k from 1 by 1 to 5 do
   kappa[k]:=sum('(-1)^j*mu[j]*(mu[1]^(k-j))*binomial(k,j)','j'=0..k):
   kappa[k]:=eval(kappa[k]):
end do:
```

Then we solve inductively for the noncentered moments, thereby expressing them as functions of the centered ones.

```
for k from 2 by 1 to 5 do
   mu[k]:=[solve(kappa[k]=s[k],mu[k])][1];
end do:
```

Finally we do the substitution and arrange the terms by powers of β.

```
for i from 1 to vdim do
   tail[i,1]:=collect(collect(simplify(
                            expand(tail[i,1])),beta),mu[1]);
end do;
```

Bibliography

J. Abate, G.L. Choudhury, D.M. Lucantoni, W. Whitt (1995). Asymptotic analysis of tail probabilities based on computation of moments, *Ann. Appl. Probab.*, 5, 983–1007.

J. Abate, G.L. Choudhury, W. Whitt (1994). Waiting-time tail probabilities in queues with long-tail service-time distribution, *Queueing Systems Theory Appl.*, 16, 311–338.

J. Abate, W. Whitt (1997). Asymptotics for M/G/1 low-priority waiting-time tail probability, *Queueing Systems Theory Appl.*, 25, 173–233.

M. Aigner (1979). *Combinatorial Theory*, Springer, 1979.

M.F. Atiyah, I.G. MacDonald (1969). *Introduction to Commutative Algebra*, Westview Press.

K.B. Athreya, P. Ney (1972). *Branching Processes*, Springer.

Ph. Barbe, W.P. McCormick (2004). Tail calculus with remainder, applications to tail expansions for infinite order moving averages, randomly stopped sums, and related topics, *Extremes*, 7, 337–365.

Ph. Barbe, W.P. McCormick (2005). Asymptotic expansions of convolutions of regularly varying distributions, *J. Austral Math. Soc., Ser. A*, 78, 339–371.

J. Beirlant, J.L. Teugels, P. Vynckier (1996). *Practical Analysis of Extreme Values*, Leuven University Press, Leuven, Belgium.

N.H. Bingham, C.M. Goldie, J.L. Teugels (1989) *Regular Variation*, 2nd ed., Cambridge

R. Bojanic, J. Karamata (1963). On slowly varying functions and asymptotic relations, Math. Research Center Tech. Report, 432, Madison, Wisconsin.

A.A. Borovkov, K.A. Borovkov (2003). On large deviation probabilities for random walks, I, regularly varying distribution tails, *Theory Probab. Appl.*, 46, 193–213.

P.J. Brockwell, R.A. Davis (1991). *Time Series: Theory and Methods*, 2nd ed., Springer.

M. Broniatowski, A. Fuchs (1995). Tauberian theorems, Chernoff inequality, and the tail behavior of finite convolutions of distribution functions, *Adv. Math.*, 116, 12–33.

J. Chover, P. Ney, S. Wainger (1973). Functions of probability measures, *J. Analyse Math.*, 26, 255–302.

Y.S. Chow, H. Teicher (1978). *Probability Theory, Independence, Interchangeability, Martingales*, Springer.

D.B.H. Cline (1986). Convolution tails, product tails and domain of attraction, *Probab. Theory Rel. Fields*, 72, 529–557.

D.B.H. Cline (1987). Convolutions of distributions with exponential and subexponential tails, *J. Austral. Math. Soc., ser. A*, 43, 347–365 (corr. *J. Austral. Math. Soc., ser A*, 48, 152–153, 1990).

J.W. Cohen (1972). On the tail of the stationary waiting-time distribution and limit theorem for M/G/1 queue, *Ann. Inst. H. Poincaré,B*, 8, 255–263.

R.A. Davis, M. Rosenblatt (1991). Parameter estimation for some time series models without contiguity, *Statist. Prob. Lett.*, 11, 515–521.

J. Delsarte (1938). Sur une extension de la formule de Taylor, *Journ. Math. Pures et Appl.*, 28(3), 213–231.

P. Diaconis, D. Freedman (1999). Iterated random functions, *SIAM Review*, 41, 45–70.

R.M. Dudley (1989). *Real Analysis and Probability*, Chapman & Hall.

P. Embrechts, C.M. Goldie, N. Veraverbeke (1979). Subexponentiality and infinite divisibility, *Z. Wahrsch. verw. Geb.*, 49, 335–347.

P. Embrechts, C. Klüppelberg, T. Mikosch (1997). *Modelling Extremal Events*, Springer.

W. Feller (1971). *An Introduction to Probability Theory and Its Applications*, vol. 2, Wiley.

J.L. Geluk (1992). Second order tail behaviour of a subordinated probability distribution, *Stoch. Proc. Appl.*, 40, 325–337.

J.L. Geluk (1994). Asymptotic behaviour of the convolution tail of distributions each having a first or second order regularly varying tail, *Analysis*, 14, 163–183.

J.L. Geluk (1996). Tails of subordinated laws: the regular varying case, *Stoch. Proc. Appl.*, 61, 147–161.

J.L. Geluk, L. de Haan, S. Resnick, C. Stărică (1997). Second-order regular variation, convolution and the central limit theorem, *Stoch. Proc. Appl.*, 69, 139–159.

J. Geluk, A.G. Pakes (1991). Second order subexponential distributions, *J. Austral. Math. Soc., Ser. A*, 51, 73–97.

J. Geluk, L. Peng, C.G. De Vries (2000). Convolutions of heavy-tailed random variables and applications to portfolio diversification and MA(1) time series, *Adv. Appl. Prob.*, 32, 1011–1026.

C.M. Goldie (1991). Implicit renewal theory and tails of solutions of random equations, *Ann. Appl. Probab.*, 1, 126–166.

D.R. Grey (1994). Regular variation in the tail behavior of solutions of random difference equations, *Ann. Probab.*, 4, 169–183.

A.K. Grincevičius (1975). On limit distribution for a random walk on the line, *Lithuanian Mat. J.*, 15, 580–589.

R. Grübel (1987). On subordinated distributions and generalized renewal measures, *Ann. Probab.*, 15, 394–415.

P. Hall, I. Weissman (1997). On the estimation of extreme tail probabilities, *Ann. Statist.*, 25, 1311–1326.

B.M. Hill (1973). A simple general approach to inference about the tail of a distribution, *Ann. Statist.*, 3, 1163–1174.

H. Kesten (1973). Random difference equations and renewal theory for product of random matrices, *Acta. Math.*, 131, 207–248.

V.P. Leonov, A.N. Shiryaev (1959). On a method of calculation of semi-invariants, *Theor. Probab. Appl.*, 3, 319–329.

P. Lévy (1954). *Théorie de l'Addition des Variables Aléatoires*, reprinted by Gabay, Paris.

L. Mattner (2004). Cumulants are universal homomorphisms into Hausdorff groups, *Probab. Theor. Rel. Fields*, 130, 151–166.

V. Marić (2000). *Regular Variation and Differential Equations*, Lecture Notes in Mathematics, 1726, Springer.

V. Marić, M. Tomić (1977). Regular variation and asymptotic properties of solutions of nonlinear differential equations, *Publ. Inst. Math. (Beograd)*, 21 (35), 119–129.

A. Nijenhuis, H. Wilf (1978). *Combinatorial Algorithms*, Academic Press, second edition.

F.W.J. Olver (1974). *Asymptotic and Special Functions*, Academic Press.

E. Omey (1981). Regular variation and its applications to second order linear differential equations, *Bull. Soc. Math. Belg.*, 32, 207–229.

E. Omey (1988). Asymptotic properties of convolution products of functions, *Publ. Inst. Math. (Beograd) (N.S.)*, 43, 41–57.

E. Omey, E. Willekens (1986). Second order behavior of the tail of a subordinated probability distribution, *Stoch. Proc. Appl.*, 21, 339–353.

E. Omey, E. Willekens (1987). Second order behavior of distributions subordinated to a distribution with finite mean, *Comm. Statist. Stoch. Models*, 3, 311–342.

A.G. Pakes (1975). On the tails of waiting-time distributions, *J. Appl. Probab.*, 12, 555-564.

A.G. Pakes (2004). Convolution equivalence and infinite divisibility, *J. Appl. Probab.*, 41, 407–424.

S. Resnick (1986). Point processes, regular variation and weak convergence, *Adv. Appl. Probab.*, 18, 66–138.

S. Resnick (1987). *Extreme Values, Regular Variation, and Point Processes*, Springer.

S. Resnick, C. Stărică (1997). Asymptotic behavior of Hill's estimator for autoregressive data, heavy tails and highly volatile phenomena, *Comm. Statist., Stoch. Models*, 13, 703–721.

S. Resnick, E. Willekens (1991). Moving averages with random coefficients and random coefficients autoregressive models, *Comm. Statist. Stoch. Models*, 7, 511–525.

T. Rolski, H. Schmidli, V. Schmidt, J. Teugels (1999). *Stochastic Processes for Insurance and Finance*, Wiley.

B. Solomyak (1995). On the random series $\sum \pm \lambda^n$ (an Erdös problem), *Ann. Math.*, 142, 611–625.

R. Stanley (1999). *Enumerative Combinatorics*, Cambridge University Press.

D. Stanton, D. White (1986). *Constructive Combinatorics*, Springer.

W. Whitt (2002). *Stochastic-Process Limits. An Introduction to Stochastic Process Limits and Their Applications to Queues*, Springer.

E. Willekens (1989). Asymptotic approximation of compound distributions and some applications, *Bull. Soc. Math. Belg., ser. B*, 41, 55-61.

E. Willekens, J.L. Teugels (1992). Asymptotic expansions for waiting time probabilities in an M/G/1 queue with long-tailed service time, *Queueing Systems Theory Appl.*, 10, 295–311.

G.E. Willmot, X.S. Lin (2000). *Lundberg Approximations for Compound Distributions with Insurance Applications*, Lecture Notes in Statistics, 156, Springer.

Editorial Information

To be published in the *Memoirs*, a paper must be correct, new, nontrivial, and significant. Further, it must be well written and of interest to a substantial number of mathematicians. Piecemeal results, such as an inconclusive step toward an unproved major theorem or a minor variation on a known result, are in general not acceptable for publication.

Papers appearing in *Memoirs* are generally at least 80 and not more than 200 published pages in length. Papers less than 80 or more than 200 published pages require the approval of the Managing Editor of the Transactions/Memoirs Editorial Board.

As of September 30, 2008, the backlog for this journal was approximately 15 volumes. This estimate is the result of dividing the number of manuscripts for this journal in the Providence office that have not yet gone to the printer on the above date by the average number of monographs per volume over the previous twelve months, reduced by the number of volumes published in four months (the time necessary for preparing a volume for the printer). (There are 6 volumes per year, each usually containing at least 4 numbers.)

A Consent to Publish and Copyright Agreement is required before a paper will be published in the *Memoirs*. After a paper is accepted for publication, the Providence office will send a Consent to Publish and Copyright Agreement to all authors of the paper. By submitting a paper to the *Memoirs*, authors certify that the results have not been submitted to nor are they under consideration for publication by another journal, conference proceedings, or similar publication.

Information for Authors

Memoirs are printed from camera copy fully prepared by the author. This means that the finished book will look exactly like the copy submitted.

Initial submission. The AMS uses Centralized Manuscript Processing for initial submissions. Authors should submit a PDF file using the Initial Manuscript Submission form found at www.ams.org/peer-review-submission, or send one copy of the manuscript to the following address: Centralized Manuscript Processing, MEMOIRS OF THE AMS, 201 Charles Street, Providence, RI 02904-2294 USA. If a paper copy is being forwarded to the AMS, indicate that it is for it Memoirs and include the name of the corresponding author, contact information such as email address or mailing address, and the name of an appropriate Editor to review the paper (see the list of Editors below).

The paper must contain a *descriptive title* and an *abstract* that summarizes the article in language suitable for workers in the general field (algebra, analysis, etc.). The *descriptive title* should be short, but informative; useless or vague phrases such as "some remarks about" or "concerning" should be avoided. The *abstract* should be at least one complete sentence, and at most 300 words. Included with the footnotes to the paper should be the 2000 *Mathematics Subject Classification* representing the primary and secondary subjects of the article. The classifications are accessible from www.ams.org/msc/. The list of classifications is also available in print starting with the 1999 annual index of *Mathematical Reviews*. The Mathematics Subject Classification footnote may be followed by a list of *key words and phrases* describing the subject matter of the article and taken from it. Journal abbreviations used in bibliographies are listed in the latest *Mathematical Reviews* annual index. The series abbreviations are also accessible from www.ams.org/msnhtml/serials.pdf. To help in preparing and verifying references, the AMS offers MR Lookup, a Reference Tool for Linking, at www.ams.org/mrlookup/.

Electronically prepared manuscripts. The AMS encourages electronically prepared manuscripts, with a strong preference for $\mathcal{A}_{\mathcal{M}}\mathcal{S}$-LaTeX. To this end, the Society has prepared $\mathcal{A}_{\mathcal{M}}\mathcal{S}$-LaTeX author packages for each AMS publication. Author packages include instructions for preparing electronic manuscripts, samples, and a style file that generates the particular design specifications of that publication series. Though $\mathcal{AMS}$-LaTeX is the highly preferred format of TeX, author packages are also available in $\mathcal{AMS}$-TeX.

Authors may retrieve an author package for *Memoirs of the AMS* from www.ams.org/journals/memo/memoauthorpac.html or via FTP to ftp.ams.org (login as anonymous, enter username as password, and type cd pub/author-info). The *AMS Author Handbook* and the *Instruction Manual* are available in PDF format from the author package link. The author package can also be obtained free of charge by sending email to tech-support@ams.org (Internet) or from the Publication Division, American Mathematical Society, 201 Charles St., Providence, RI 02904-2294, USA. When requesting an author package, please specify $\mathcal{AMS}$-LaTeX or $\mathcal{AMS}$-TeX and the publication in which your paper will appear. Please be sure to include your complete mailing address.

After acceptance. The final version of the electronic file should be sent to the Providence office (this includes any TeX source file, any graphics files, and the DVI or PostScript file) immediately after the paper has been accepted for publication.

Before sending the source file, be sure you have proofread your paper carefully. The files you send must be the EXACT files used to generate the proof copy that was accepted for publication. For all publications, authors are required to send a printed copy of their paper, which exactly matches the copy approved for publication, along with any graphics that will appear in the paper.

Accepted electronically prepared files can be submitted via the web at www.ams.org/submit-book-journal/, sent via FTP, or sent on CD-Rom or diskette to the Electronic Prepress Department, American Mathematical Society, 201 Charles Street, Providence, RI 02904-2294 USA. TeX source files, DVI files, and PostScript files can be transferred over the Internet by FTP to the Internet node ftp.ams.org (130.44.1.100). When sending a manuscript electronically via CD-Rom or diskette, please be sure to include a message identifying the paper as a Memoir.

Electronically prepared manuscripts can also be sent via email to pub-submit@ams.org (Internet). In order to send files via email, they must be encoded properly. (DVI files are binary and PostScript files tend to be very large.)

Electronic graphics. Comprehensive instructions on preparing graphics are available at www.ams.org/authors/journals.html. A few of the major requirements are given here.

Submit files for graphics as EPS (Encapsulated PostScript) files. This includes graphics originated via a graphics application as well as scanned photographs or other computer-generated images. If this is not possible, TIFF files are acceptable as long as they can be opened in Adobe Photoshop or Illustrator. No matter what method was used to produce the graphic, it is necessary to provide a paper copy to the AMS.

Authors using graphics packages for the creation of electronic art should also avoid the use of any lines thinner than 0.5 points in width. Many graphics packages allow the user to specify a "hairline" for a very thin line. Hairlines often look acceptable when proofed on a typical laser printer. However, when produced on a high-resolution laser imagesetter, hairlines become nearly invisible and will be lost entirely in the final printing process.

Screens should be set to values between 15% and 85%. Screens which fall outside of this range are too light or too dark to print correctly. Variations of screens within a graphic should be no less than 10%.

Inquiries. Any inquiries concerning a paper that has been accepted for publication should be sent to memo-query@ams.org or directly to the Electronic Prepress Department, American Mathematical Society, 201 Charles St., Providence, RI 02904-2294 USA.

Editors

This journal is designed particularly for long research papers, normally at least 80 pages in length, and groups of cognate papers in pure and applied mathematics. Papers intended for publication in the *Memoirs* should be addressed to one of the following editors. The AMS uses Centralized Manuscript Processing for initial submissions to AMS journals. Authors should follow instructions listed on the Initial Submission page found at www.ams.org/memo/memosubmit.html.

Algebra to ALEXANDER KLESHCHEV, Department of Mathematics, University of Oregon, Eugene, OR 97403-1222; email: ams@noether.uoregon.edu

Algebraic geometry and its application to MINA TEICHER, Emmy Noether Research Institute for Mathematics, Bar-Ilan University, Ramat-Gan 52900, Israel; email: teicher@macs.biu.ac.il

Algebraic geometry to DAN ABRAMOVICH, Department of Mathematics, Brown University, Box 1917, Providence, RI 02912; email: amsedit@math.brown.edu

Algebraic topology to ALEJANDRO ADEM, Department of Mathematics, University of British Columbia, Room 121, 1984 Mathematics Road, Vancouver, British Columbia, Canada V6T 1Z2; email: adem@math.ubc.ca

Combinatorics to JOHN R. STEMBRIDGE, Department of Mathematics, University of Michigan, Ann Arbor, Michigan 48109-1109; email: FRS@umich.edu

Complex analysis and harmonic analysis to ALEXANDER NAGEL, Department of Mathematics, University of Wisconsin, 480 Lincoln Drive, Madison, WI 53706-1313; email: nagel@math.wisc.edu

Differential geometry and global analysis to LISA C. JEFFREY, Department of Mathematics, University of Toronto, 100 St. George St., Toronto, ON Canada M5S 3G3; email: jeffrey@math.toronto.edu

Dynamical systems and ergodic theory and complex anaysis to YUNPING JIANG, Department of Mathematics, CUNY Queens College and Graduate Center, 65-30 Kissena Blvd., Flushing, NY 11367; email: Yunping.Jiang@qc.cuny.edu

Functional analysis and operator algebras to DIMITRI SHLYAKHTENKO, Department of Mathematics, University of California, Los Angeles, CA 90095; email: shlyakht@math.ucla.edu

Geometric analysis to WILLIAM P. MINICOZZI II, Department of Mathematics, Johns Hopkins University, 3400 N. Charles St., Baltimore, MD 21218; email: trans@math.jhu.edu

Geometric analysis to MARK FEIGHN, Math Department, Rutgers University, Newark, NJ 07102; email: feighn@andromeda.rutgers.edu

Harmonic analysis, representation theory, and Lie theory to ROBERT J. STANTON, Department of Mathematics, The Ohio State University, 231 West 18th Avenue, Columbus, OH 43210-1174; email: stanton@math.ohio-state.edu

Logic to STEFFEN LEMPP, Department of Mathematics, University of Wisconsin, 480 Lincoln Drive, Madison, Wisconsin 53706-1388; email: lempp@math.wisc.edu

Number theory to JONATHAN ROGAWSKI, Department of Mathematics, University of California, Los Angeles, CA 90095; email: jonr@math.ucla.edu

Partial differential equations to GUSTAVO PONCE, Department of Mathematics, South Hall, Room 6607, University of California, Santa Barbara, CA 93106; email: ponce@math.ucsb.edu

Partial differential equations and dynamical systems to PETER POLACIK, School of Mathematics, University of Minnesota, Minneapolis, MN 55455; email: polacik@math.umn.edu

Probability and statistics to RICHARD BASS, Department of Mathematics, University of Connecticut, Storrs, CT 06269-3009; email: bass@math.uconn.edu

Real analysis and partial differential equations to DANIEL TATARU, Department of Mathematics, University of California, Berkeley, Berkeley, CA 94720; email: tataru@math.berkeley.edu

All other communications to the editors should be addressed to the Managing Editor, ROBERT GURALNICK, Department of Mathematics, University of Southern California, Los Angeles, CA 90089-1113; email: guralnic@math.usc.edu.

Titles in This Series

923 **Michael Jöllenbeck and Volkmar Welker,** Minimal resolutions via algebraic discrete Morse theory, 2009

922 **Ph. Barbe and W. P. McCormick,** Asymptotic expansions for infinite weighted convolutions of heavy tail distributions and applications, 2009

921 **Thomas Lehmkuhl,** Compactification of the Drinfeld modular surfaces, 2009

920 **Georgia Benkart, Thomas Gregory, and Alexander Premet,** The recognition theorem for graded Lie algebras in prime characteristic, 2009

919 **Roelof W. Bruggeman and Roberto J. Miatello,** Sum formula for SL_2 over a totally real number field, 2009

918 **Jonathan Brundan and Alexander Kleshchev,** Representations of shifted Yangians and finite W-algebras, 2008

917 **Salah-Eldin A. Mohammed, Tusheng Zhang, and Huaizhong Zhao,** The stable manifold theorem for semilinear stochastic evolution equations and stochastic partial differential equations, 2008

916 **Yoshikata Kida,** The mapping class group from the viewpoint of measure equivalence theory, 2008

915 **Sergiu Aizicovici, Nikolaos S. Papageorgiou, and Vasile Staicu,** Degree theory for operators of monotone type and nonlinear elliptic equations with inequality constraints, 2008

914 **E. Shargorodsky and J. F. Toland,** Bernoulli free-boundary problems, 2008

913 **Ethan Akin, Joseph Auslander, and Eli Glasner,** The topological dynamics of Ellis actions, 2008

912 **Igor Chueshov and Irena Lasiecka,** Long-time behavior of second order evolution equations with nonlinear damping, 2008

911 **John Locker,** Eigenvalues and completeness for regular and simply irregular two-point differential operators, 2008

910 **Joel Friedman,** A proof of Alon's second eigenvalue conjecture and related problems, 2008

909 **Cameron McA. Gordon and Ying-Qing Wu,** Toroidal Dehn fillings on hyperbolic 3-manifolds, 2008

908 **J.-L. Waldspurger,** L'endoscopie tordue n'est pas si tordue, 2008

907 **Yuanhua Wang and Fei Xu,** Spinor genera in characteristic 2, 2008

906 **Raphaël S. Ponge,** Heisenberg calculus and spectral theory of hypoelliptic operators on Heisenberg manifolds, 2008

905 **Dominic Verity,** Complicial sets characterising the simplicial nerves of strict ω-categories, 2008

904 **William M. Goldman and Eugene Z. Xia,** Rank one Higgs bundles and representations of fundamental groups of Riemann surfaces, 2008

903 **Gail Letzter,** Invariant differential operators for quantum symmetric spaces, 2008

902 **Bertrand Toën and Gabriele Vezzosi,** Homotopical algebraic geometry II: Geometric stacks and applications, 2008

901 **Ron Donagi and Tony Pantev (with an appendix by Dmitry Arinkin),** Torus fibrations, gerbes, and duality, 2008

900 **Wolfgang Bertram,** Differential geometry, Lie groups and symmetric spaces over general base fields and rings, 2008

899 **Piotr Hajłasz, Tadeusz Iwaniec, Jan Malý, and Jani Onninen,** Weakly differentiable mappings between manifolds, 2008

898 **John Rognes,** Galois extensions of structured ring spectra/Stably dualizable groups, 2008

897 **Michael I. Ganzburg,** Limit theorems of polynomial approximation with exponential weights, 2008

TITLES IN THIS SERIES

896 **Michael Kapovich, Bernhard Leeb, and John J. Millson,** The generalized triangle inequalities in symmetric spaces and buildings with applications to algebra, 2008

895 **Steffen Roch,** Finite sections of band-dominated operators, 2008

894 **Martin Dindoš,** Hardy spaces and potential theory on C^1 domains in Riemannian manifolds, 2008

893 **Tadeusz Iwaniec and Gaven Martin,** The Beltrami Equation, 2008

892 **Jim Agler, John Harland, and Benjamin J. Raphael,** Classical function theory, operator dilation theory, and machine computation on multiply-connected domains, 2008

891 **John H. Hubbard and Peter Papadopol,** Newton's method applied to two quadratic equations in $\mathbb{C}^2$ viewed as a global dynamical system, 2008

890 **Steven Dale Cutkosky,** Toroidalization of dominant morphisms of 3-folds, 2007

889 **Michael Sever,** Distribution solutions of nonlinear systems of conservation laws, 2007

888 **Roger Chalkley,** Basic global relative invariants for nonlinear differential equations, 2007

887 **Charlotte Wahl,** Noncommutative Maslov index and eta-forms, 2007

886 **Robert M. Guralnick and John Shareshian,** Symmetric and alternating groups as monodromy groups of Riemann surfaces I: Generic covers and covers with many branch points, 2007

885 **Jae Choon Cha,** The structure of the rational concordance group of knots, 2007

884 **Dan Haran, Moshe Jarden, and Florian Pop,** Projective group structures as absolute Galois structures with block approximation, 2007

883 **Apostolos Beligiannis and Idun Reiten,** Homological and homotopical aspects of torsion theories, 2007

882 **Lars Inge Hedberg and Yuri Netrusov,** An axiomatic approach to function spaces, spec tral synthesis and Luzin approximation, 2007

881 **Tao Mei,** Operator valued Hardy spaces, 2007

880 **Bruce C. Berndt, Geumlan Choi, Youn-Seo Choi, Heekyoung Hahn, Boon Pin Yeap, Ae Ja Yee, Hamza Yesilyurt, and Jinhee Yi,** Ramanujan's forty identities for Rogers-Ramanujan functions, 2007

879 **O. García-Prada, P. B. Gothen, and V. Muñoz,** Betti numbers of the moduli space of rank 3 parabolic Higgs bundles, 2007

878 **Alessandra Celletti and Luigi Chierchia,** KAM stability and celestial mechanics, 2007

877 **María J. Carro, José A. Raposo, and Javier Soria,** Recent developments in the theory of Lorentz spaces and weighted inequalities, 2007

876 **Gabriel Debs and Jean Saint Raymond,** Borel liftings of Borel sets: Some decidable and undecidable statements, 2007

875 **C. Krattenthaler and T. Rivoal,** Hypergéométrie et fonction zêta de Riemann, 2007

874 **Sonia Natale,** Semisolvability of semisimple Hopf algebras of low dimension, 2007

873 **A. J. Duncan,** Exponential genus problems in one-relator products of groups, 2007

872 **Anthony V. Geramita, Tadahito Harima, Juan C. Migliore, and Yong Su Shin,** The Hilbert function of a level algebra, 2007

871 **Pascal Auscher,** On necessary and sufficient conditions for L^p-estimates of Riesz transforms associated to elliptic operators on $\mathbb{R}^n$ and related estimates, 2007

870 **Takuro Mochizuki,** Asymptotic behaviour of tame harmonic bundles and an application to pure twistor D-modules, Part 2, 2007

869 **Takuro Mochizuki,** Asymptotic behaviour of tame harmonic bundles and an application to pure twistor D-modules, Part 1, 2007

For a complete list of titles in this series, visit the
AMS Bookstore at **www.ams.org/bookstore/**.